The Evolution of the Universe

Richard Lewis

ISBN: 9798872468233

CONTENTS

1 INTRODUCTION

Why we embark on a line of investigation is sometimes a mystery. Perhaps it is a basic human need to make sense of the world. Sometimes it is the irresistible challenge of solving problems. It might also be that sense of wishing to leave some legacy for future generations.

Whatever the reason, I found myself becoming engaged in the great quest to understand the evolution of the universe. Throughout most of my adult life the Big Bang theory of the origin of the universe has been the dominant cosmology. The Big Bang theory won out against the Steady State theory back in the 1960s and has been developed and refined and modelled in great detail.

Yet what is it about the theory that leaves you with an uncomfortable feeling that it is just not quite right? When faced with a theory such as the Big Bang there is a process of evaluation to be sure that it should truly be adopted as part of a personal world view. As we learn about the world around us we gradually build up a mental model which we adjust based on new information. It is important that our world view is correct in every aspect so that our actions are based on a firm foundation.

In order to understand the evolution of the universe it is necessary to discard the Big Bang theory. Prior to the 1960s there were competing theories of the evolution of the universe namely the Steady State theory and the Big Bang theory. After the discovery of the Cosmic Microwave Background Radiation (CMBR) a scientific consensus formed around the Big Bang theory. It was argued that the CMBR was an echo of the Big Bang. We will come back to challenge this assumption later but for now let's discuss the problems that developed arising from the Big Bang theory.

In order to confirm physical theories it is necessary to create mathematical models and for this purpose it was necessary to make some physical assumptions. One such assumption was the cosmological principle which can be stated as: "In modern physical cosmology, the cosmological principle is the notion that the spatial distribution of matter in the universe is equally distributed and isotropic when viewed on a large enough scale, since the forces are expected to act equally throughout the universe on a large scale, and should, therefore, produce no observable inequalities in the large-scale structuring over the course of evolution of the matter field that was initially laid down by the Big Bang."

From this starting assumption it was possible to derive equations which describe the evolution of the universe in a mathematical model. The cosmological principle admits the possibility that the universe is infinite or finite with no boundary. The possibility that the universe is finite with a space boundary is not compatible with the cosmological principle.

From the mathematical models it is then possible to trace back in time towards the supposed Big Bang event and to assess the mathematical model against observed reality. Over the years it has been necessary to introduce adjustments to the model to be compatible with observations. It was necessary to introduce the idea of inflation to account for the uniformity in the observed temperature in different directions. This idea of inflation required a rapid exponential expansion in the early seconds of the universe which subsequently reduced to a slower rate of expansion.

Then observations started to show that there is considerably more matter in the universe than that detectable through observed radiation. This so called dark matter was detected through its gravitational effect on the rotation of galaxies and through the gravitational lensing of light passing around galaxies. The measurements suggest around 85% of matter is dark matter and 15% is luminous matter. Attempts were made to explain dark matter using different laws of gravitation (MOND) but these ideas were not widely accepted because they contradict the general theory of relativity which is a well evidenced theory. So the existence of dark matter is accepted but without any description of the matter particle for dark matter.

The mathematical model is given the name lambda CDM where CDM stands for Cold Dark Matter and lambda refers to the cosmological constant. This cosmological constant is included as an additional term in the Einstein equations of general relativity to account for dark energy. More on dark energy later.

So even though the lambda CDM Big Bang model seemed rather contrived, most cosmologists were content for several decades as it seemed to match observations reasonably well. Difficulties started to arise when observations suggested that there were stars, globular clusters and galaxies which must have formed more than 13.8 billion years ago. There are also observations which give conflicting results for the expansion of the universe and this is known as the Hubble tension.

In the history of physics and cosmology it seems that the evolution of ideas does not follow a steady path but from time to time there is a fundamental change of ideas. The prevailing theory (in this case the Big Bang) is refined and refined until suddenly it is contradicted. It is the case that it is not logically possible to prove that a physical theory is correct but it is possible to prove that a physical theory is false. We are at that point now with the Big Bang theory.

Once our mindset moves from trying to understand the Big Bang theory to trying to challenge and find fault with the big Bang theory our analysis follows a different path.

The first idea within the Big Bang theory that did not make sense was the formation of galaxies. It just did not seem correct that the gravitational attraction of a vast dispersed cloud of gas and dust would lead to the isolated and distinctive galaxy formations that we observe.

The story of how galaxies form from a dispersed cloud of gas and dust just does not make logical sense. Imagine a dispersed cloud of gas and dust moving under the effect of gravitational attraction. We can imagine local points of increased density leading to the formation of a star. We can then imagine several stars moving under gravitational attraction to form a galaxy.

However, if we think about an individual star, there will be cases where the distribution of stars around that individual star will result in neutral gravity so we should see isolated stars in between galaxies in at least some cases. We don't see this and all galaxies appear to be formed from the centre outwards. There is a super massive black hole at the centre surrounded by a star region which in turn is surrounded by a dark matter halo.

The most important aspect of the Big Bang theory to challenge is the explanation for the cause of the Cosmic Microwave Background Radiation (CMBR). The physical explanation for the CMBR is recombination when the universe suddenly goes from opaque to transparent as the temperature drops through 3000 degrees Kelvin. Then the radiation is released so the CMBR starts. But in order to be observing the CMBR with a look back time of 13.8 billion years (since the Big Bang supposedly happened everywhere) the radiation must then stop.

So it is suggested that we are observing a flash of radiation which is an "echo" of the Big Bang. It is to be noted that the temperature of 3000 degrees comes from a calculation of the temperature that would be needed to result in the observed temperature of 2.7 degrees Kelvin today. Thus the temperature of 3000 degrees is derived from the model and not from some laboratory measurement of a "recombination" process involving a plasma. So the whole Big Bang hypothesis rests on an area of plasma physics which has not been verified in the laboratory.

This casts serious doubt around the whole argument which resulted in the adoption of the Big Bang theory.

Once a fundamental theory is adopted it requires extraordinary evidence to displace that theory. Cosmology is built one step at a time and every new step once adopted becomes part of the standard model of cosmology. The model will be fine tuned and adjusted and new observations will result in new hypotheses for dark matter, dark energy, inflation and so on.

What it takes is some amazing new technology such as the James Webb Space Telescope (JWST) to show us that the theoretical models are wrong. The JWST is showing us galaxies that are perfectly formed and must have begun their development more than 13.8 billion years ago.

It might be helpful to state clearly at this stage the assumptions behind the Big Bang theory which I believe to be false:

1. The assumption that all matter formed very early in the life of the universe

2. The cosmological principle

3. The assumption regarding the cause of the Cosmic Microwave Background radiation

The starting hypothesis for the Space Boundary theory is that the universe is finite with a space boundary. Starting from a spherical region of empty space with a space boundary the universe expands due to the presence of the space boundary. Tension builds up in the fabric of space until energy is released in a galaxy formation event quite close to the centre of the finite universe. Subsequently galaxies form so that the number of galaxies in the universe increases by a factor of about 20 every 14 billion years. Galaxy formation takes place over a wide range of distance with new galaxy formation overlapping the existing galaxy distribution.

The rest of the model is a logical progression from this starting point. The Space Boundary theory has some aspects of the Big Bang theory (the expansion of space) and some aspects of the Steady State theory. The Steady State theory assumed that the universe is infinite and galaxies form everywhere in the space between galaxies. The Space Boundary theory asserts that galaxies formed starting around 126 billion years ago in a finite universe with a boundary and the position of galaxy formation moved progressively from the centre towards the boundary.

So how should I present the theory in a way that it can be most easily understood. I find that an audiovisual slide presentation provides the best method of conveying understanding. I have created a number of such presentations on YouTube and I have provided in this book the text from selected presentations along with the slides. The Youtube presentations are identified and freely available online.

There are two presentations provided in Chapter 2 and 3. The first is titled "The Explanation for Dark Matter and Dark Energy" and was prepared as preparation for an online conference where the conference paper of the same title was first published.

The second presentation titled "The Formation of the Solar System" describes the formation of the planets from matter which came from the fast rotating sun.

The text is transcribed from the video with corrections and clarification in brackets.

The conference paper titled "The Explanation for Dark Matter and Dark Energy" is presented in Chapter 4. The preprint titled "The Evolution of the Universe" is provided in Chapter 5. This is the source document on which the conference paper is based.

Finally in Chapter 6 there are references to the YouTube videos on the subject which describe particular aspects of the model and the underlying justification.

In the first three chapters of the book the narrative is descriptive with limited mathematical analysis and this should be sufficient to understand the Space Boundary theory. The subsequent chapters are aimed at anyone wishing to verify the theory and it contains additional mathematical analysis.

The proposed list of five books presented in Chapter 7 describes the scope of the Open World initiative and this book is the first in the series.

2 THE EXPLANATION FOR DARK MATTER AND DARK ENERGY

The Explanation for Dark Matter and Dark Energy

ICDMDEPS 2020: 14. International Conference on Dark Matter and Dark Energy in Physical Sciences

November 05-06, 2020 in Amsterdam, Netherlands

Author: Richard Lewis

This presentation has been prepared as a pre-presentation for the International Conference on Dark Matter and Dark Energy, taking place in Amsterdam on November the 5th and 6th, and the topic of this presentation is the Explanation for Dark Matter and Dark Energy.

In this paper, presented at the conference, I will be explaining that to get to the bottom of this mystery about dark matter and dark energy, we actually have to change our fundamental understanding of the evolution of the universe.

In fact, we have to discard the Big Bang Theory and look for an alternative. Now, what first convinced me that the Big Bang Theory was wrong was looking, about 15 years ago, I found that the data from NASA was generally available. The availability of galaxy position data could be downloaded by anyone. You didn't need to be a researcher. You could just go online to the NASA Extragalactic Database and download the data for the positions of galaxies.

As a personal interest I did this and I thought it would be interesting to look at the distribution for patterns and what I found was that the distribution of galaxies was completely inconsistent with the Big Bang theory. In the sense that, if you look at the number of galaxies by distance, you find that there's a very tight grouping for the evolution of this curve, and convinced me that there had to be an alternative to the Big Bang theory.

Slide 1 - Why should we look for alternatives to the Big Bang theory?

- Dark matter and Dark energy unexplained
- The Big Bang theory does not conform to the law of conservation of energy
- Observations of objects older than 13.8 billion years

So why should we look for an alternative? Well, firstly, we've been unable to explain dark matter and dark energy. Then, it's clear that the Big Bang theory does not conform to the law of conservation of energy. And more and more, with more ability to analyse what we're seeing in the universe, it's becoming clear that there are objects such as stars and globular clusters which are older than 13.8 billion years, which is the age of the universe as determined by the Big Bang theory. So it makes no sense to have objects within the universe which are older than the universe itself.

Slide 2 - The false assumptions of the Big Bang theory

- That all matter formation took place at the same time
- The cosmological principle that the universe is homogeneous and isotropic
- The explanation for the Cosmic Microwave Background Radiation

So having developed an alternative to the Big Bang theory, which I will be presenting here in this presentation, it then becomes clear which of the assumptions of the Big Bang theory are in fact false. They're listed here, the three assumptions which are not correct. The point to emphasise here is that these are assumptions.

Now, the Big Bang theory says we look at all these galaxies and we go back in time, and the galaxies get closer and closer together, and the universe gets hotter and hotter. And the implicit assumption is that all that matter, when we go back in time, all that matter formed at one time.

The cosmological principle is that we're not in any special place in the universe. We've been caught out like this before. We thought we were special and that the universe looks the same wherever you are. It's got the same physical properties and looks the same in all directions. The physical properties are the same, but it turns out, we are in a rather particular place quite close to the centre of a finite universe, which I'll come to.

Then, the explanation for the cosmic microwave background radiation which is the key element which convinced everyone that the Big Bang theory was right and the Steady State theory was wrong was the Cosmic Microwave Background Radiation, which it's claimed is the echo of the Big Bang coming from recombination.

Slide 3 - The Hot Big Bang

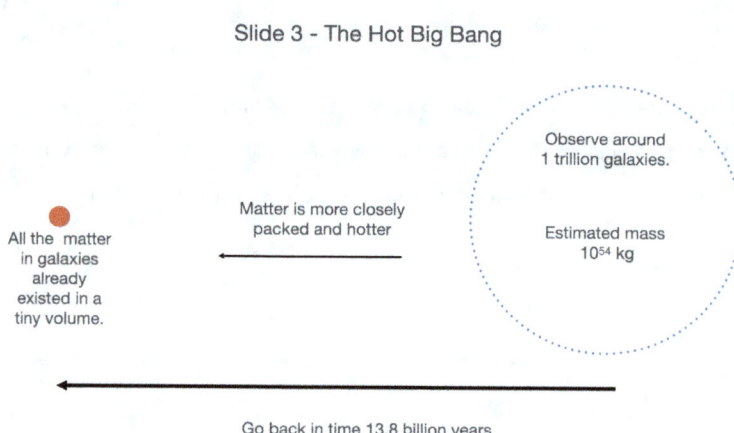

So this slide shows an illustration of the hot Big Bang model. And you can see on the right, we've got today, inside the blue circle, we've got around a trillion galaxies with an estimated mass of 10 to the 54 kilograms (10^{54} kg). So we're talking about, 10 to the 12 galaxies (10^{12} galaxies), each of which is around 10 to the 42 (10^{42} kg) in mass. And that mass, the Hot Big Bang model says you take that back in time, and all that matter already exists in a tiny volume the size of a grapefruit.

- Is the universe infinite?
  - An infinite universe requires infinite energy
- If finite is there a space boundary?
  - Finite with no boundary has been a popular theory but would require the universe to be curved back on itself
- Start with the assumption of a universe which is finite with a space boundary

So if we don't like the Big Bang theory, and if we've got grave doubts about that, let's go back to very first principles and start thinking about what the nature of the universe would be and there are actually three possibilities, in terms of finite and infinite. So, we could assume that the universe was infinite, or we could assume that it's finite, without a boundary.

So these first two points are what people would usually argue between. Some say its infinite, some say it's finite, but with no boundary. What this proposal is saying is that we have to look at the case where we think of the universe as finite with a space boundary. For a very good reason. An infinite universe would require infinite energy.

If the universe were finite, with no boundary, then that means it curves back on itself and studies of the cosmic microwave background radiation show that in the radial direction, the universe is completely flat. So that rules out that possibility.

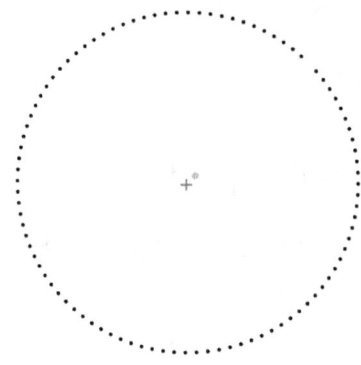

- A finite, bounded universe has a centre

- The Milky Way galaxy is moving at 552km/s relative to the CMB rest frame

- Using Hubble's law we are around 26 million light years from the centre of the universe

- This explains why the universe appears similar in all directions

- The cosmological principle is a false assumption

So, here in slide five, we can, once we've taken the point that we have a universe which is finite with a boundary, that implies that, the universe has a centre.

And it also implies a rest frame.

We have observed the rest frame, which is the Cosmic Microwave Background Rest Frame. And we can measure the speed of the Milky Way galaxy relative to this rest frame. And using Hubble's Law, we can then fix our position relative to the centre of the universe.

So you can think of this model of the universe as being a space where all galaxies are moving away from the centre of the universe due to the expansion of space, and individually each galaxy will also move under whatever gravitational effect it is subjected to.

From using this observational data, we find that the Milky Way galaxy is around 26 million light years from the centre of the universe, which puts us quite close to the centre of the galaxy distribution, and this explains why the galaxy distribution seems similar in all directions and explains why we might have come to the conclusion that the cosmological principle is true. But in fact, in a finite universe the cosmological principle must be false because the position close to the boundary is clearly different from the position close to the centre.

- Matter formation uses energy released from expanding spacetime

- Matter formation happened progressively over billions of years

- Total Energy is conserved (and equal to zero) when you define total energy to include mass, energy and spacetime curvature

- $(8\pi G/c_4)\ T_{\mu\nu} + \frac{1}{2}\ g_{\mu\nu}\ R - R_{\mu\nu} = 0$

- Einstein equation rearranged means mass + energy + spacetime curvature = zero

So, we said in the Big Bang theory that it violates the law of conservation of energy. So let's think about how we can develop a theory which conforms to the law of conservation of energy. Our thoughts about conservation of energy have changed over the years. We used to think that mass was conserved. We used to think of energy being conserved, and we thought of these as being completely independent.

But with E equals m c squared ($E = mc^2$), it's clear that mass energy is conserved, and mass can be converted into energy.

What this slide tells us, and what general relativity tells us, is that you have to generalise again, your conservation of energy law to make total energy equal to mass plus energy plus spacetime curvature. And this equation that I'm showing here is just the Einstein equations reorganised a little bit to say mass plus energy plus spacetime curvature is zero. And in this equation, mass is always positive, energy is always positive, and [the energy associated with] spacetime curvature is always negative. So with this in mind, you can see that to balance the energy equation, the spacetime curvature contribution has the ability to provide the energy for matter formation.

Slide 7 - Energy for Galaxy Formation

- Starting from a finite region of empty space with a space boundary
- Assume the universe expands at a constant rate
- To balance the total energy equation, matter formation is needed
- This happens as a result of galaxy formation events

So we'll start with the, the simplest idea. We'll start with a spherical region of empty space with a space boundary. We'll assume that the universe expands at a constant rate. And we can measure this rate directly because we can observe the expansion of the universe and measure it fairly accurately.

Slide 8 - The Expansion of the Universe

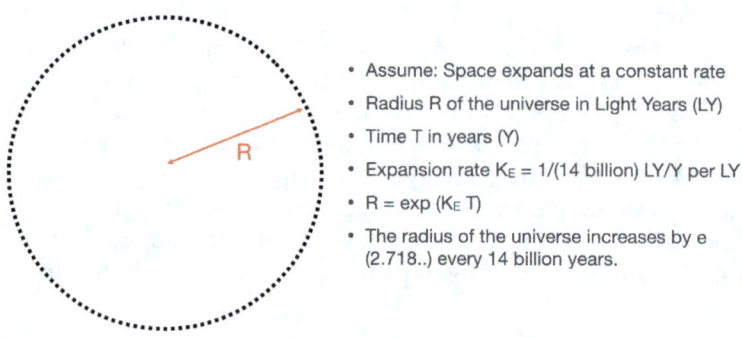

- Assume: Space expands at a constant rate
- Radius R of the universe in Light Years (LY)
- Time T in years (Y)
- Expansion rate K_E = 1/(14 billion) LY/Y per LY
- R = exp (K_E T)
- The radius of the universe increases by e (2.718..) every 14 billion years.

I tend to use an expansion rate, which is one part in 14 billion light years per year, per light year and that would correspond to 69.84 kilometres per second per Mega parsec, which is the units that, astronomers tend to use. And in this model, to balance the total [energy] equation we need matter formation. And the matter formation is not at some initial beginning of things, it happens as a result of galaxy formation events over an extended period of time.

Slide eight. So now we're going to look at the expansion of the universe. And using some fairly straightforward mathematics, it becomes clear that if we're assuming that the expansion of every segment of space is the same, then if we take the radius of the universe, the expansion of the universe by simple integration produces an exponential relationship between the radius R and the time T.

I find it very convenient to work with R in light years and the time in years. And from this formula, it tells us that the radius of the universe increases by a factor of around 2.7 every 14 billion years.

Slide 9 - Matter formation in the Universe

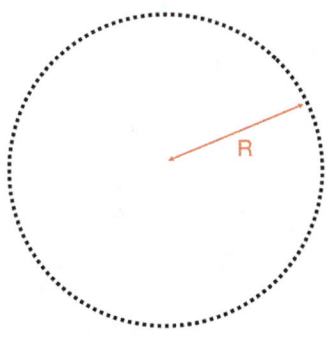

- Assume: Total mass in the universe is proportional to the volume of the universe

- Average matter density in the universe remains constant at less than 1 hydrogen atom per cubic meter

- The volume of the universe increases by e^3 (~20) every 14 billion years.

- The total mass in the universe increases by a factor of 20 every 14 billion years

- The total number of galaxies increases by a factor of 20 every 14 billion years

- Estimated time since the first galaxy formed is >126 billion years based on 20^9 galaxies

[In slide 9, delete the second bullet point on average matter density] So let's think a bit more about this process of matter formation in the universe. We're going to make a couple of assumptions here, which seem reasonable that the matter generation capacity of each unit volume in the universe is the same. So the total mass in the universe, we can assume will be proportional to the total volume of the universe.

15

We can tell from observations that the average matter density in the universe [that we might observe] is something of the order of one hydrogen atom per cubic metre. [The average density required to create an event horizon at 8.77 billion light years is around 14 hydrogen atoms per cubic metre. The very approximate estimate of one hydrogen atom per cubic metre is based on galaxy counts out to a distance of 500 million light years. This observation is consistent with the model because the majority of galaxies are moving away from the centre of the universe towards the event horizon. This reduces the matter density as observed locally.]

Now if the volume of the universe increases by e to the power three (e³) so we're talking about a volume expansion derived from a linear expansion. So we cube the expansion rate to get a volume expansion of around 20, a factor of 20 every 14 billion years. So this tells us that the total mass in the universe in the form of galaxies will increase by a factor of 20 every 14 billion years.

So this gives us an estimated time since the first galaxy formed of 126 [billion years], which is 14 times 9. And that number 9 comes from the observation that we're taking a rough ballpark estimate of 20 to the power 9 galaxies, which is [approximately] 500 billion. So these are just ballpark figures, but you can see that we're going way beyond the 13.8 billion years age of the universe that is currently proposed.

Slide 10 - The position of galaxy formation

Galaxies initially form near the centre of the universe.
The predominant position of galaxy formation moves outwards towards the boundary

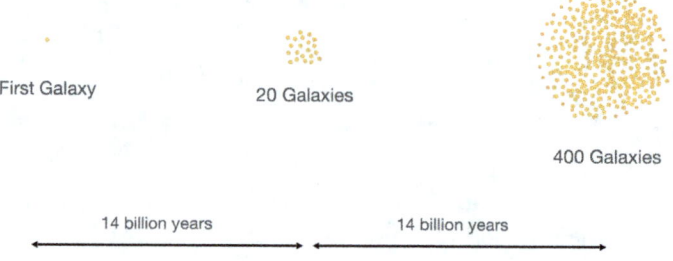

First Galaxy
20 Galaxies
400 Galaxies

14 billion years   14 billion years

This is how the position of galaxy formation evolves. We have in slide 10, the first galaxy forms. 14 billion years later, we have 20 galaxies. 14 billion years after that, we have 400 galaxies. And so on.

Now the position of galaxy formation moves outwards towards the boundary, and in fact, most of the galaxy formation today is going on beyond 14 billion [light] years, we can deduce this from the model. And that explains why we don't see directly a large amount of current galaxy formation.

What we do see, very occasionally, at distances of 6, 10 billion light years is huge gamma ray bursts, which no one can believe are a single multidirectional event. They would assume that it must be a jet because it's so energetic, it outshines the entire universe. And what those distant gamma ray bursts are, I would claim, is a galaxy formation event because the energy that's being released is of the order of 10 to the 42 kilograms (10^{42} kg) to form the mass of the new galaxy and what we're observing in the gamma rays is the energy that has not been converted into matter.

Slide 11 - Cosmic Microwave Background Radiation (CMBR)

- The CMBR is radiation that has travelled for 13.8 billion years
- The source of the radiation must have been at 8.77 billion light years from the centre of the universe at a time 13.8 billion years ago
- There was sufficient mass in the universe to create an event horizon of radius 8.77 billion light years at a time 13.8 billion years ago
- The CMBR is radiation from a distribution of galaxies moved outwards by the expansion of the universe but constrained by the event horizon

So we need to tackle this thorny question of what is the cosmic microwave background radiation because It can't be what the Big Bang theory thinks it is, which is an echo of the Big Bang.

By analysing this problem, it's very clear that you can use the Schwarzschild solution to estimate the existence of an event horizon at a distance which must have been at 8.77 billion light years from the centre at a time 13.8 billion years ago.

Now, we've been used to using the Schwarzschild radius calculation for black holes and indeed, though it proves very accurate, and we always think of an event horizon as being viewed from the outside, or not viewed at all because it's a black hole, we don't see anything. But the mathematical equations show that the event horizon effectively produces a demarcation or a threshold or a boundary, in a way, between the region inside the event horizon and the region outside the event horizon.

But the region inside the event horizon is space time and the equations are general, and even though we've previously applied them to black holes, which are a very small, dense distribution, there's absolutely no reason why the same equations can't be applied to a distribution of matter, which is very low density, which results in a very large Schwarzschild radius.

It turns out that the density required to produce an event horizon at 8.77 billion light years is about 0.724 hydrogen atoms per cubic metre. [This figure of 0.724 hydrogen atoms per cubic metre is a prediction of the density that we might observe in the local galactic region. It assumes that the matter density of around 14 hydrogen atoms per cubic metre needed to form the event horizon is affected by the subsequent expansion of space over 13.8 billion years.]

Slide 12 - The Event Horizon, Observation Horizon and Space Boundary

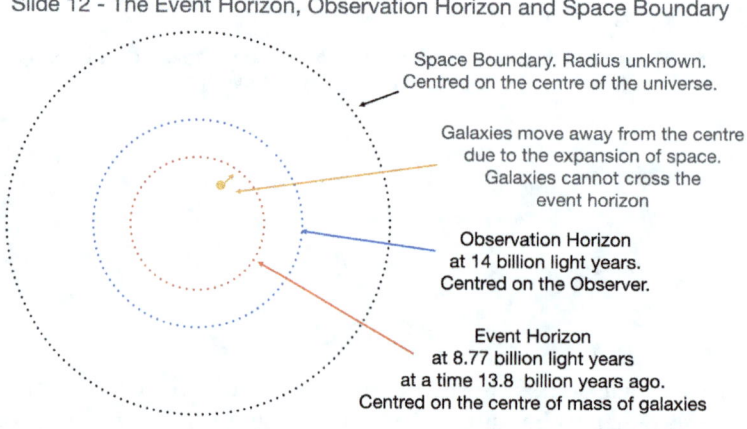

Space Boundary. Radius unknown.
Centred on the centre of the universe.

Galaxies move away from the centre
due to the expansion of space.
Galaxies cannot cross the
event horizon

Observation Horizon
at 14 billion light years.
Centred on the Observer.

Event Horizon
at 8.77 billion light years
at a time 13.8 billion years ago.
Centred on the centre of mass of galaxies

So here's an illustration of the various horizons and boundaries that we understand. So the space boundary is shown as the outer black dotted circle in this diagram. [This is where there is an obvious requirement to address the question of what is outside the space boundary. The best answer I have to this is that there is no outside and the entire universe resides within the boundary. It is a similar question to asking what happened before the Big Bang. The answer to the question of what happens when matter or radiation tries to cross the space boundary is to be found in Open World book 2 - The nature of matter.]

The observation horizon is at a distance of 14 billion [light] years centred upon the observer, but because we as an observer are quite close to the centre of the universe we can take all of these circles as being concentric. And finally, the event horizon is, or I should say was, at 8.77 billion light years at a time 13.8 billion years ago.

And you can see that light coming from just within the event horizon, even though it was at 8.77 billion light years would take 13.8 billion years to reach us because of the expansion of space and this this relationship between the time and the expansion it can be calculated and this is referenced in the paper.

So what I believe is happening is that the galaxy formation starts close to the centre, moves outwards and eventually there's enough galaxies to form an event horizon and at that point any galaxies within the event horizon will be subject to the expansion of space, but they certainly can't cross the event horizon. [The Schwarzschild radius calculation shows that it would require around 28 billion galaxies to form an event horizon at 8.77 billion light years].

So the hypothesis is that there will be a buildup of galaxies, almost like a shell of galaxies, just inside the event horizon by this process. And the proposal then, and this needs a lot of work to convince me and everyone else that this is the correct assumption that the cosmic microwave background radiation is coming from this distant field of galaxies and over time, the temperature of the radiation will cool from the heat of the galaxy down to the observed cosmic microwave background.

So although I say I'm not one hundred percent sure of the exact source, what is clear is that there will be an event horizon there and the cosmic microwave background is consistent with radiation coming from a source just inside the event horizon. [It is also possible that cosmic ray particles created during galaxy formation events are also held at the event horizon and emit radiation contributing to the CMBR.]

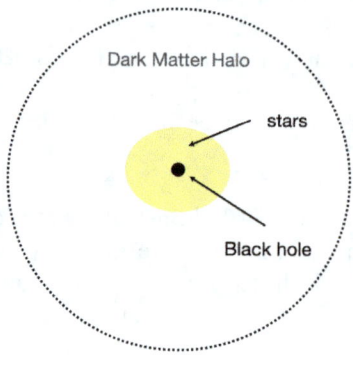

- Galaxy formation event creates neutrons
- Pairs of neutrons bond to form di-neutrons which are the dark matter particle
- Single neutrons decay to form a proton and an electron and an antineutrino
- Protons and electrons form hydrogen atoms which form hydrogen molecules
- Protons and electrons and di-neutrons form helium atoms
- Gravitational acceleration towards the centre forms a super massive neutron star
- A super massive neutron star is observed as a super massive black hole

So now, now we come down to, following all this preamble, we've been saying, well, we're going to explain dark matter, we're going to explain dark energy. Let's get down to that topic. So the proposal is that a galaxy formation event, the release of energy from space time, creates a huge wave disturbance. And ultimately, there are very few stable particles. What is formed, in fact, perhaps from a spiral wave, a series of neutrons with a shared axis of rotation. [This explanation of the formation of matter in a galaxy formation event will make more sense after reading Open World book 2 - The nature of matter]

And these can pair up and bond to form dineutrons, and that's the proposed dark matter particle.

So, there we have the search for the dark matter particle is over, and many people have said, well, it couldn't be baryonic, because we've looked at all of the particles of the standard model and none of them meet the criteria. But no one has actually sat down and said, well, maybe it's a pair of particles or a bonded particle. And that's really what is the proposed solution.

[It might be the case that dineutrons produced in a galaxy formation event are in a different state to the dineutrons that we might produce in a laboratory on Earth. However, it will be the case that when the hydrogen and dark matter particles form stars that the dark matter particles decay to two neutrons. This would explain the reason why dark matter is not to be found in the solar system because it forms part of the matter included in the formation of stars.]

So, from the neutrons, we get dark matter and single neutrons. Some don't pair, and before they can pair, they form a proton and an electron through the normal decay process that we're familiar with. That leads to hydrogen atoms and hydrogen molecules and you can also see that if you had a dineutron which captured a proton, or in fact, captured two protons, then you'd find you've got a nucleus of two protons and two neutrons, and then an outer shell of two electrons, and you've got a helium atom.

So this can explain the production of helium right at the galaxy formation event. And then all of this matter can accelerate towards the centre [of the galaxy]. At the same time, the whole halo, the whole spherical region is expanding partly from the explosion and partly from the expansion of space.

So the gravitational acceleration affects mostly the matter close to the centre, which collapses down to the centre and typically one thousandth of the total mass of the galaxy can condense into the centre in the form of a massive neutron star, which we observe as a black hole.

Slide 14 - Dark Energy

- Dark energy does not exist and the expansion of the universe is not accelerating.
- Dark energy is the wrong explanation for the unexpected recession velocity of distant galaxies
- Galaxy recession velocities have two components:
  1. Due to the expansion of space
  2. Due to movement through space from gravitational acceleration
- Distant galaxies have a greater velocity component towards the centre of the universe due to gravitational acceleration

Finally, we come to dark energy, and we have to really go back to what was observed, because what was observed and led to the concept of dark energy was the unexpected recession velocity of distant galaxies. And because of that, the conclusion was taken, well, in that case, we're looking at some galaxies which were there many, many years into the past and therefore the universe expansion must be accelerating.

But that's not the case. What's happening is that these distant galaxies are affected by the gravitational acceleration due to the distribution of galaxies located closer to the centre of the universe. So as well as an expansion component there is an [gravitational] acceleration towards the centre of the universe and this component is what is causing the unexpected recession velocity of the distant galaxies.

So, dark energy is not real, the expansion of space is not accelerating and the effect is caused by our misunderstanding of the fundamentals of cosmology.

3 THE FORMATION OF THE SOLAR SYSTEM

Today I am going to be talking about the formation of the solar system and we are going to start from the formation of the Milky Way galaxy.

The Formation of the Milky Way Galaxy

- The Milky Way galaxy is located around 26 million light years from the centre of the universe

- The Milky Way galaxy is towards the outer part of the 400 galaxies closest to the centre of the universe

- The Milky Way galaxy formed around 28 billion years after the first galaxy formed about 126 billion years ago

- The Milky Way galaxy formed an estimated 100 billion years ago

Now based on a new model of the evolution of the universe in which the first galaxy formed around 126 billion years ago, we can look at the position of the Milky Way galaxy and the galaxy is located around 26 million light years from the centre of the universe. Based on the evolutionary model, the number galaxies to form every 14 billion years increases by a factor of 20. So the first galaxy formed, then 14 billion years later there were 20 galaxies, a further 14 billion years later that was increased to 400. We can look at the region between the Milky Way galaxy and the centre of the universe and we conclude that we are in the outer part of the 400 galaxies closest to the centre of the universe.

So that means that the Milky Way galaxy must have formed around 28 billion years after the first galaxy. So that puts the formation of the Milky Way galaxy an estimated 100 billion years ago. [A revised model of galaxy formation suggests that the formation process is more spread out over distance. The position of a galaxy is not a reliable guide to its age. The estimated time of formation of the Milky Way galaxy is 56 to 70 billion years ago.]

Spiral Galaxy Formation

- Spiral galaxies form from the merger of two spherical regions of gas and dust

- A galaxy from the first merger is a spiral galaxy with two spiral arms

- The structure of the Milky Way galaxy suggests it resulted from a merger of two spiral galaxies - a merger of a merger

- Galaxy mergers create star formation regions

- The heavier elements are formed in a supernova

Now the Milky Way galaxy didn't form in its present form at that time. What formed then were spherical regions of gas and dust and what we can model is that if two spherical regions of gas and dust form in the vicinity of each other, what happens is that the tidal forces create the spiral arms and we get from that a spiral galaxy with two spiral arms.

Now the structure of the Milky Way galaxy suggests that it results from not just a simple structure of two spiral arms but from a merger of two spiral galaxies.

Now this process of galaxy merger takes typically 20 to 30 billion years so for a merger of a merger we are talking around 40 to 60 billion years.

Angular Momentum

- Any rotating object in the universe acquired its angular momentum from the merger of galaxies

- This applies to the rotation of spiral galaxies, the rotation of black holes and the rotation of the solar system

- Angular momentum is an important factor in determining the outcome of a star formation event

- A small component of angular momentum results in a single star

- A large component of angular momentum results in a binary

When the spiral galaxy forms in the first instance it creates star forming regions and in the process of evolution once the stars have formed we get the heavier elements from a supernova explosion.
An important factor to take into account when considering the evolution of galaxies and star systems is angular momentum because angular momentum is conserved.

It is conserved over a merger and it is conserved over a supernova explosion. So we can say that any rotating object in the universe acquired its angular momentum from the merger of two galaxies.

This applies to the reason why spiral galaxies are rotating, and the reason why the central black holes are rotating often at quite high speeds, and the rotation of the solar system the angular momentum for that actually comes down to an effect from the merger of galaxies.

When you look down at the level of a star, you can see that when a star forms from a diffuse cloud of elements, gas, dust a small component of angular momentum will result in a single star but a large component of angular momentum must result in a binary system because that is the only way in which that larger quantity of angular momentum can be accommodated.

We find that a third of the star systems in the Milky Way galaxy are either binary or multiple star systems and it's not obvious when you look into the

Nearly a binary system

- About one third of the star systems in the Milky Way galaxy are binary or multiple.

- The solar system can be thought of as nearly a binary with the sun and Jupiter as the potential stars

- There was not enough angular momentum for a binary system and too much angular momentum for a single stable star

- Hypothesis: The solar system formed initially as a single fast rotating star which was unstable

night sky because these binaries are close together [and] they appear as a single star to the naked eye.

Now the solar system can be thought of as something that nearly formed as a binary system with the sun and Jupiter as the potential points of the two potential stars within the system.

But Jupiter is much much smaller than the sun and in fact there wasn't enough angular momentum for a binary system to form in the solar system but there was too much angular momentum for a single stable star to form.

So our hypothesis is that the solar system formed initially as a single fast rotating star which was unstable.

If we look at the way stars form, they form from a cloud of gas and dust and heavy elements the cloud of gas and dust that formed the solar system must already have contained all the elements of the periodic table which are

Composition of cloud of gas and dust

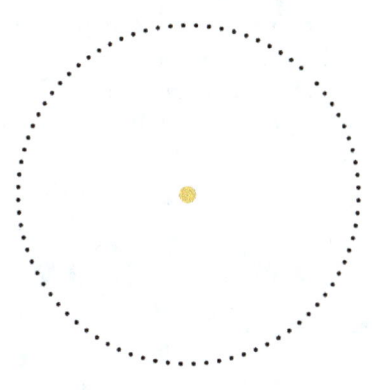

- Heavier elements formed in supernova explosion

- They might have been several generations of star formation and supernova

- The cloud contained all the elements of the periodic table found on Earth

- The cloud formed from a star which was rotating prior to supernova

found on Earth. That cloud which formed from a star was rotating because that is the observation that the solar system is rotating.

So if you imagine, we said earlier that a rotating star which goes supernova will preserve its angular momentum so if this cloud of matter originated from a supernova, the matter would have been ejected out by the supernova explosion but the trajectory if we take a single point the trajectory of the

Matter collapses to the centre

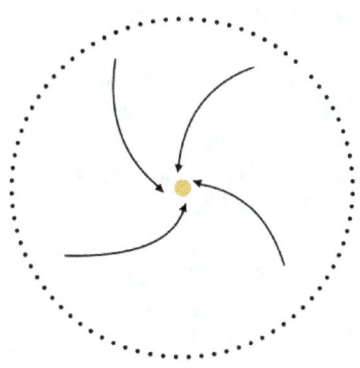

matter would be to go out into a kind of orbital path but ultimately it will collapse back into the central region and that will form a star.

So what this says if you imagine a point - a particle here towards the periphery of the cloud of gas and dust there is no way that that particle could fall in towards the centre and then go into some kind of protoplanetary disk of gas and dust in orbit around the sun because its trajectory must be back into the centre.

Speed of rotation increases - side view

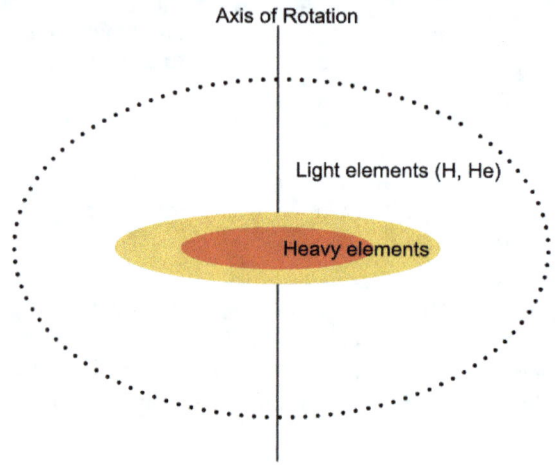

So this rules out the current idea that the planets formed from a protoplanetary [disk of] gas and dust in orbit.

So what the hypothesis says is that the material in the cloud collapsed towards the centre and it's rotating so if we imagine the heavier elements falling towards the centre and the lighter elements like hydrogen and helium being at the outer region of the cloud you can see that because of the rotation, the shape is no longer spherical.

It turns into a shape where the profile is an ellipse and if you think about what happens to an ice skater who is rotating on the ice if they pull their arms into the centre they rotate faster.

Speed of rotation increases - top view

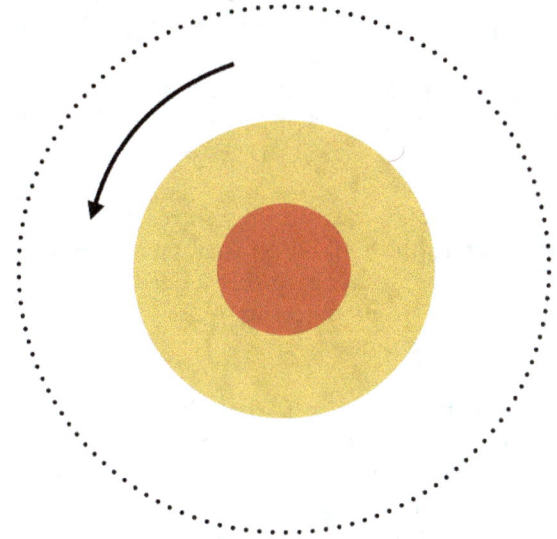

This is the mechanism that is happening as the elements fall in towards the centre, the core of the system rotates faster and faster.

Star becomes unstable

- Star material is thrown out from the sun with high angular momentum
- Rotating mass spins off planets and their moons into orbit around the sun
- Much additional debris in the form of asteroids is released
- Cleared by the planets leaving just the asteroid belt and the Kuiper belt
- A substantial part of the rotating mass falls back into the sun
- Only planets with circular orbits survive
- The oxygen layer combined with the abundant hydrogen forms water

So that speed of rotation increases and looking from the top it's still circular but because of the shape it becomes increasingly unstable.

The hypothesis is that lack of stability means that a body of material is ejected from the sun and if this were just a simple body of matter you might argue that this is going to be projected out and then it's going to fall back in towards the sun on an orbital style path with the sun as a focus.

So all the matter would fall back into the centre. But this doesn't take into account the fact that this [the sun] is fast rotating and in particular the body of material [ejected] is fast rotating so this indeed would spin off material in all directions as it goes on its trajectory out as far as the Kuiper belt and it's releasing material all the time.

That process of rotating spins off the planets and their moons and additional debris which is in the form of the asteroids. But the important point is that when we say planets are thrown off in all directions only those planets with a circular orbit will survive because a planet on an elliptical orbit will fall back towards the sun and eventually be recaptured by the sun.

So what's left from this chaotic start, what's left is a well ordered system of planets in an approximately circular motion.

Indeed the presence of planets in their orbit will clear out any asteroids that happen to have been left in that vicinity. So this is why we see all the craters on the moon because the asteroids which were in the Earth belt have fallen in to the Earth and the moon.

Of course any asteroids falling on Earth will over time due to the nature of the Earth most traces would be lost.

Now we talked about the heavier elements falling in towards the centre so if you can imagine the structure as being a layered structure of elements, approximately in the order of the periodic table as they collapse in, and so somewhere within this structure we have an oxygen layer so that when the core, the core is rotating faster than the outer regions, the core becomes unstable and breaks through the outer layers taking the oxygen with it into the hydrogen layer.

So we have got a sudden mixing of oxygen and hydrogen as part of this process and this then explains the abundance of water.

It's quite reasonable to suppose that from our observation that a lot of that

Evidence

- The formation of the planets and their moons

- The rotation of the Earth and the orbit of the moon

- The hot molten core of the Earth

- The presence of water

- The asteroid belts

- The presence of helium on Earth

water layer appears to have formed in the orbital region of the Earth. Not necessarily part of the Earth in its initial formation but gradually accreting onto the Earth because individual molecules of water can still be in orbit out in space.

So what evidence do we have for this hypothesis? Well it's hard to see how a protoplanetary disk could form both the planets and the moons in orbit and with the rotation they have.

The hot molten core of the Earth is material coming from the hot molten core of the sun and it's hard to imagine that that hot molten core would be able to be accreted by a process of just cold dust particles in a protoplanetary disk.

We talked about the presence of water. The asteroid belt; well the asteroids don't look like the accretion of gas and dust, they look like fragments of solidified liquid material from this process which we are describing.

The whole idea of accretion of tiny particles into larger particles, because gravity on that scale is very weak, the idea that the accretion of particles could form asteroids and planets doesn't really make sense.

Finally the presence of helium on Earth. If everything accreted based on a cloud of gas and dust we can see that the process of the formation of the planets would have the same source material. So the Earth would have been surrounded by hydrogen and helium but that would have all escaped the Earth's atmosphere because the gravity on Earth is not enough to hold hydrogen and helium in place. But we do find helium on Earth and we find

Pale blue dot

it in large caves in various places and this is where the chaotic formation of the molten Earth had bubbles of hydrogen and helium which were trapped and are still here to this day. That process couldn't have happened if it accreted from gas and dust because the helium would just have dissipated.

So that's a long history. It goes back a hundred billion years and I mention the wonderful picture that was taken of planet Earth from a very great distance and it just reveals the Earth to be a pale blue dot in space. So thinking of this very very long history to create a habitable planet with life we need to think about sustainability and how we ensure that this habitat, this cradle, is sustainable not just for ten years, not just for a hundred years but ongoing.

4 THE EXPLANATION FOR DARK MATTER AND DARK ENERGY - CONFERENCE PAPER

Abstract—The following assumptions of the Big Bang theory are challenged and found to be false: the cosmological principle, the assumption that all matter formed at the same time and the assumption regarding the cause of the cosmic microwave background radiation. The evolution of the universe is described based on the conclusion that the universe is finite with a space boundary. This conclusion is reached by ruling out the possibility of an infinite universe or a universe which is finite with no boundary. In a finite universe, the centre of the universe can be located with reference to our home galaxy (The Milky Way) using the speed relative to the Cosmic Microwave Background (CMB) rest frame and Hubble's law. This places our home galaxy at a distance of approximately 26 million light years from the centre of the universe. Because we are making observations from a point relatively close to the centre of the universe, the universe appears to be isotropic and homogeneous but this is not the case. The CMB is coming from a source located within the event horizon of the universe. There is sufficient mass in the universe to create an event horizon at the Schwarzschild radius. Galaxies form over time due to the energy released by the expansion of space. Conservation of energy must consider total energy which is mass (+ve) plus energy (+ve) plus spacetime curvature (-ve) so that the total energy of the universe is always zero. The predominant position of galaxy formation moves over time from the centre of the universe towards the boundary so that today the majority of new galaxy formation is taking place beyond our horizon of observation at 14 billion light years.

Keywords—Cosmic microwave background, dark energy, dark matter, evolution of the universe.

I.   The Space Boundary Theory

THE approach taken is to set aside the Big Bang theory and try to deduce the evolution of the universe from different starting assumptions. The Big Bang theory deduces the evolution of the universe by assuming the cosmological principle [1]-[8] and a uniform progress back in time with a corresponding increase in temperature and density. However, evolution moves forward in time and it is entirely possible that a different evolution path leads to the universe that we observe today. The starting configuration of the universe for the Space Boundary theory is a spherical region of space with a space boundary which then expands at a fixed rate which is the same rate as is observed today. This puts the age of the universe at more than 330 billion years so we have to find a different explanation for the cosmic microwave background radiation (CMBR) [9]-[13].

The Big Bang theory describes the CMBR as radiation arising from the Big Bang at a time 13.8 billion years ago [14]-[17]. The Space Boundary theory describes the CMBR as radiation coming from a source located within the event horizon of the universe. This removes the time constraint on the age of the universe and allows the assumption of uniform expansion to be possible. This requires a discussion about the event horizon of the universe. The idea of an event horizon arises directly from the equations of the General Theory of Relativity [18]-[21]. Where there is a region of space with sufficient mass lying entirely within that region, spacetime is curved to such an extent that nothing can escape. Any high energy particles or radiation generated within this region of space will not be able to escape through the event horizon.

The condition that we observe radiation coming from the vicinity of the event horizon with a travel time of 13.8 billion years is that the radiation must have originated from a distance of 8.77 billion light years. Using this Schwarzschild radius it is possible to calculate the amount of mass in the form of galaxies that would be required to cause this event horizon. Then taking that mass density forward in time to the present, it gives a matter density consistent with current observational data. So the universe that we observe is consistent with the existence of an event horizon causing the CMBR.

II.   A Finite Universe

The conclusion is reached by ruling out the possibility of an infinite universe or a universe which is finite with no boundary. An infinite universe would require infinite energy for matter formation and would violate the law of conservation of energy.

A finite universe with no boundary is ruled out by the observation that the radiation from the CMBR follows a straight line path meaning that on the large scale of the universe the curvature is flat in the radial direction to the CMBR [22].

Given that the universe is finite with a boundary and making the symmetrical assumption that the boundary is spherical this means that the universe has a centre.

The space within the boundary of the universe is a spherical region of space which is expanding away from the centre of the universe and the recession velocity of each point of space is proportional to the distance from the centre of the universe.

III.   Our Position in the Universe

The CMBR provides a useful frame of reference for the universe called the CMB rest frame. This means that there will be a point located at the centre of the universe which is at rest with reference to the CMB rest frame.

When the CMBR is measured in great detail it reveals that the Home galaxy (the Milky Way) is moving at a speed of 552 km/sec with reference to the CMB rest frame [23], [24].

The universe is expanding uniformly in all directions at a steady rate of 1 part in 14 billion light years per year per light year [25]-[29]. Each galaxy in the universe can be thought of as being embedded in this expanding frame of reference where the recession velocity of each galaxy depends on the distance from the centre of the universe.

Using the velocity of the Home galaxy and the rate of expansion of space and using Hubble's law [30]-[32] we can locate the centre of the universe at a distance of approximately 26 million light years from the Home galaxy. Put another way, this means that the Home galaxy is located 26 million light years from the centre of the universe and is moving away from the centre of the universe at 552 km/sec.

IV.   Galaxy Formation

Going back 200 billion years, there were no galaxies, just an expanding spherical region of space and then galaxy formation started an estimated 126 billion years ago with the source of energy for the formation of galaxies is the energy released from the expansion of space and the galaxy formation events start closest to the centre of the universe.

The predominant position of formation of new galaxies moved away from the centre of the universe towards the boundary so that by now the majority of galaxy formation takes place beyond the observation horizon at 14 billion light years.

A galaxy formation event creates a spherical region in which the release of energy from space leads to the formation of neutrons some of which immediately bond to form neutron groups of two or more neutrons bonded together. These neutron groups are what we observe as dark matter [33]-[38].

Single neutrons will decay over a short time period to create a proton an electron and a neutrino. The protons and electrons can together form hydrogen atoms and then hydrogen molecules.

Some of the material (hydrogen and dark matter) will fall under gravity towards the central region of the galaxy and result in the formation of a large neutron star which will be viewed externally as a super massive black hole.

V.   Dark Energy

The existence of dark energy [39], [40] has been proposed to explain the variation in the recession velocity of more distant galaxies. The recession velocity of galaxies was expected to follow Hubble's law where the recession velocity of a distant object is proportional to its distance. The recession velocity of a distant galaxy is calculated by measuring the red shift of the light coming from the galaxy.

For observations of the more distant galaxies, it was found that the recession velocity for a given distance was less than expected under Hubble's law [41], [42]. This measurement was made possible by using distant supernovae to give an accurate estimate of distance using the luminosity of the supernova event. The conclusion taken from this unexpected recession velocity was that the expansion of the universe must have been slower in the past and that the expansion of the universe must be accelerating. The cause of this accelerated expansion was named dark energy. The alternative proposal is that the universe expansion is not accelerating but is uniform over time and distance. Instead, the difference in recession velocity for more distant galaxies is found by taking into consideration the gravitational acceleration directed towards the centre of the universe.

VI.   The Variation of the Radius of the Universe with Time

The proposal analysed here is that the recession velocity of the boundary from the centre of the universe is proportional to the radius. This means that to a first level of approximation, the rate of expansion of space is constant over time and distance.

T is time measure in years. R is distance measured in light years. Then $\Delta R/\Delta T = K_E R$ where K_E is a positive constant. This means that $\Delta T = (1/K_E) \Delta R/R$. Integrating this expression we get:

$$T = (1/\ K_E)\ \log\ (R) + C \qquad\qquad (1)$$

Since the time value can be set to an arbitrary reference point we take the value of $T = 0$ to be when the radius of the universe was 1 light year. This makes the constant of integration (C) zero. From (1) for T we get: $R = \exp(K_E\ T)$. Also differentiating to get the recession velocity we get: $dR/dT = K_E\ \exp(K_E\ T)$.

To obtain the value of K_E we use the observed local expansion of space as a first approximation. We also assume as a first approximation that the expansion of space is uniform out to the boundary. Currently the observation of galaxy recession velocities shows that $(dR/dT)/R$ is equal to $1/(14$ billion) which is therefore the value of K_E. (Note that the value of $1/(14$ billion) light years per year per light year is taken from observations of the value of the Hubble constant. This value of K_E corresponds to a Hubble constant value of 69.84 km/s per Mpc. If subsequent observations suggest a different value for K_E this will affect the numerical results of this paper by a few percentage points.) So we have two equations relating T and R.

$$T = (1/\ K_E)\ \log\ (R)$$

$$R = \exp(K_E\ T)$$

To get an initial estimate of the current value of T and R, the observational evidence of the CMBR is used. The CMBR is coming from the event horizon of the universe and we observe the CMBR having travelled for 13.8 billion years at the speed of light.

The position of the event horizon 13.8 billion years ago can be calculated by considering the expansion of space at one part in 14 billion light years per year per light year during the travel time.

The calculation divides the total travel time of the CMB radiation into equal intervals dt in such a way that the contribution to the distance measurement at time t is given by $dt/\exp\ (K_E\ t)$ for each time segment dt. Then by integrating $dt/\exp\ (K_E\ t)$ from $t = 0$ to $t = 13.8$ billion we obtain the result $14*(1 - \exp(-13.8/14))$ billion light years for the position of the event horizon at a time 13.8 billion years ago.

The event horizon must have been at approximately 8.775584 billion light years from the centre of the universe at the time 13.8 billion years ago. The CMB radiation has covered the original separating distance of 8.775584 billion light years during a travel time of 13.8 billion years because over that time interval the expansion of space has increased the distance to be covered.

For the CMB to be visible, the radius to the boundary must have been greater than the radius to the event horizon at a time 13.8 billion years ago. The minimum value for R at a time 13.8 billion years ago is therefore 8.77 billion light years. Using (1), T is currently greater than 320.53 + 13.8 = 334.33 billion years. Using R = exp(K_E T), the radius of the universe (R) is currently greater than 23.52 billion light years.

VII.   The Formation of Mass in the Universe

As the universe expands, the total energy (mass plus energy plus spacetime curvature) must remain the same. The expansion of space results in an increase in the radius of curvature of space at every point in space. There will be a relationship between the total mass formed within the universe and the volume of the universe. The proposed relationship is that the total mass of galaxies in the universe is proportional to the volume of the universe.

The formula M proportional to R^3 together with R = exp (K_E T) implies that every 14 billion years the mass of all the galaxies in the universe increases by a factor of e^3 = 20.0855. This new galaxy formation is currently taking place mostly beyond 14 billion light years so beyond our observation horizon. Occasionally a galaxy formation event occurs within our range of observation and then we observe a large gamma ray burst coming from 6 to 10 billion light years.

Given that the number of galaxies in the universe increases by a factor of 20 every 14 billion years we can use this to make a rough estimate of the time since the formation of the first galaxy. When the first galaxy formed there was just one galaxy in the universe. After 14 billion years there were 20 galaxies. After 28 billion years there were 400 galaxies. After 14x9 billion years there were 20^9 galaxies. This is the same as saying that after 126 billion years there were around 500 billion galaxies. Given that we can estimate the number of galaxies as greater than 500 billion, we can say that the time since the first galaxy formed is greater than 126 billion years.

VIII.   The Age of the Universe

The formulas for the expansion of space developed in Appendix 1 are:

$$T = (1/ K_{E)} \log (R)$$

$$R = \exp(K_E T)$$

where K_E = 1/14billion Light Years/Year per Light Year (LY/Y per LY)

In an elapsed time of 14 billion years the expansion of any region of space more than doubles: Suppose $T_2 = T_1 + 14$ billion years.

Then $\log(R_2) = \log(R_1) + 1$. $R_2 = R_1 \times e$ where e is the exponential constant and has a value of approximately 2.71828. So if we were to assume that the equations of expansion operated down to the smallest scales then if we put $R_1 = 1$ cm then $R_2 = 2.71828$ cm. This implies that it takes 14 billion years for the universe to expand from 1 cm to just under 3 cm. It is not safe to assume that the expansion of space at such small scales follows the uniform expansion equations.

The approach taken is to apply the expansion equation for positive values of T only, meaning that the equation $R = \exp(K_E T)$ is only used for values of R greater than 1 light year. The characteristics of the expansion of the universe for values of R of less than 1 light year require further analysis outside the scope of this paper.

The analysis of the evolution of the universe in the Space Boundary Theory then starts when $T = 0$ and $R = 1$ light year. The universe is at least 334 billion years old as calculated in Appendix 1.

The question of whether the age of the universe is finite or infinite is outside of the scope of the Space Boundary Theory.

The universe may or may not have had a beginning.

IX.   Dark Matter Analysis

During galaxy formation the initial galaxy formation event results in the formation of neutrons in numbers corresponding to the total mass of the galaxy. We know that a single neutron will decay into a proton and an electron after an average period of around 15 minutes. If two neutrons collide before they decay into a proton and an electron then the neutrons will bond into a dineutron as this is a lower energy state. We also know that neutrons in a bonded state do not decay so easily into protons and electrons. This is a similar situation to neutrons in an atomic nucleus where the decay of a neutron is a low probability. So the formation of neutron pairs or possibly higher numbers of neutrons (neutron groups) during galaxy formation would have two effects. Firstly we would expect these neutron groups to fall under gravity and form the central black hole. Secondly, neutron groups would not be detectable by photons and would pervade the galaxy halo thus being an ideal candidate for dark matter.

Any material that falls under gravity towards the centre of the galaxy will increase the mass of the super massive neutron star which we observe as a super massive black hole at the centre of each galaxy.

The idea of neutron groups as dark matter depends on the binding energy of two neutrons being positive. The binding energy corresponds to the mass defect associated with the neutron to neutron bond.

X.   Black Holes

The description of the formation of galaxies includes the formation of the super massive black hole at the centre of every galaxy. The formation of the galaxy starts with the formation of a number of neutrons equivalent to the total mass of the galaxy. A proportion of the neutrons form neutron groups and move under the effect of gravity to form the central black hole.

In this section we investigate the hypothesis that all black holes contain neutron stars and that the event horizon of the black hole is caused by the mass and high density of the neutron star within the event horizon. One prediction from this hypothesis is that we should be able to see a range of sizes of neutron stars and black holes but any black hole will always be greater in mass than the largest observed neutron star. At present the largest observed neutron star is 1.97 solar masses and the smallest observed black hole is 3.8 solar masses.

We can estimate the size of neutron star where the event horizon is exactly at the surface of the star. The following symbols are used in the calculation: N is the number of neutrons in the neutron star; m is the mass of a neutron 1.675×10^{-27} kg; r is the radius of a neutron; R is the Schwarzschild radius [43] of the star; M is the mass of the star.

Schwarzschild radius formula gives: $R = 2GM/c^2$. We can start with $M = Nm$ for the mass of the star. $R = 2GNm/c^2$ gives the radius of the star. The volume of the star is $4/3 \, \pi \, R^3$. The volume of the star is also $N \times$ volume of the neutron $= 4/3 \, \pi \, r^3$. So $4/3 \, \pi \, R^3 = N \times 4/3 \, \pi \, r^3$. Therefore $R^3 = N \, r^3$

$$(2GNm)^3 = c^6 N r^3$$

$$N^2 = c^6 r^3 / (2Gm)^3$$

Now we have to decide on the value of r to use. The effective radius value calculated in Appendix 4 of the Unification of Physics [44] is 0.630058×10^{-15} m. This gives a value of N as approximately 4.03×10^{57}. This number of neutrons has a mass of 6.751×10^{30} kg which is 3.4 solar masses. The value of R is given by $N^{1/3} r$ which is 10.027 kilometres. The minimum radius for a black hole is approximately 10 km. This lends support to the theory that all black holes contain neutron stars which cause the event horizon of the black hole.

The calculated value for the largest observable neutron star (3.4 solar masses) and the smallest possible black hole (3.4 solar masses) is consistent with observation.

For black holes larger than 3.4 solar masses, the position of the event horizon at the Schwarzschild radius increases in proportion to the mass whereas the radius of the neutron star within the event horizon increases in proportion to the cube root of the mass. So for a supermassive black hole of 3.4 million solar masses the radius of the neutron star would be approximately one thousand kilometres whereas the Schwarzschild radius would be 10 million kilometres.

In summary, the space boundary theory and the spacetime wave theory point to the conclusion that the internal structure of a black hole is not a singularity but a neutron star of a mass corresponding to the observed mass of the black hole.

As of 2017, there has been considerable success with gravitational wave detectors which detect merging black holes and merging neutron stars [44], [45]. The neutron stars observed are always less than 3.4 solar masses and the merging black holes are all greater than 3.4 solar masses. The production of gravitational waves is the same in the two cases because, in both cases, we are observing merging neutron stars.

When two black holes merge the angular momentum of the system causes a rotating black hole to form. This is in fact a merger of two neutron stars which results in a rotating neutron star inside the event horizon. Where the mass inside the event horizon is rotating the calculation of the event horizon at the Schwarzschild radius has to be modified to take into account the rotation of the mass.

The event horizon may not be closed at the axis of rotation and the jets which are characteristic of rotating neutron stars can also be visible coming from the axis of the merged black hole. This opening of the event horizon at the axis also applies to the merging of galaxies where the central black holes combine and under the right conditions of mass and rotation the jets can emerge from the central black hole of the merged galaxy along the axis of rotation.

The neutron star may also contain a proportion of protons. The proton is approximately the same mass and volume as the neutron so the above calculation for the event horizon still applies. The electrostatic repulsion of the positively charged protons is overcome by the gravitational forces of the star. In the case of a rotating neutron star with a proportion of protons, the effect of the protons is to create an electric charge moving in a circular path, which then creates a strong magnetic field.

Neutron stars vary in their total mass, their speed of rotation and the proportion of protons and this variation then affects the ability of the neutron star to generate high energy jets emerging at the two poles along the axis of rotation.

VI.   CMBR

CMBR has been mapped in detail and is the key piece of evidence in support of the Big Bang theory. In the Big Bang theory the CMBR is believed to be the radiation coming from the Big Bang itself following a process called recombination.

In the Space Boundary theory of the evolution of the universe it is proposed that the CMBR is radiation coming from the vicinity of the event horizon of the universe. The General Theory of Relativity (GR) provides equations which allow us to calculate the curvature of spacetime due to a given distribution of mass. These equations have a solution first proposed by Schwarzschild: $R = 2Gm/c^2$. This equation can be used to calculate the event horizon for a given distribution of matter. The event horizon is at the radius R for a mass m. Nothing, not even light can escape beyond the event horizon. This formula is used to calculate the event horizon for a black hole and in this case we are looking from the outside towards a black hole which we can detect because of its gravitational effect on other objects. The equation can also be applied to the distribution of mass within the universe where we are observing from within the event horizon and in this case it shows that there will be a radius from which matter and radiation cannot escape.

Based on the observation that the radiation was emitted 13.8 billion years ago we can calculate that at the time of the emission of this radiation the distance to the source was 8.77 billion light years. This implies an event horizon at a radius of 8.77 billion light years at a time 13.8 billion years ago. We can then calculate the matter density required to create an event horizon at this distance. From this we can project that matter density forward in time assuming an expansion of 1/(14 billion) light years per year per light year and we find that the matter density observed today should be around 0.724 hydrogen atoms per cubic metre. So, the analysis is consistent with the CMB radiation coming from a source in the vicinity of the event horizon. The nature of this source of radiation is still a subject of investigation but the following hypothesis is presented.

The formation of galaxies takes place at points in space which are located progressively further from the centre of the universe. The model suggests that the number of galaxies increases by a factor of 20 every 14 billion years and the average position of formation of these galaxies moves from the centre of the universe outwards towards the boundary. The actual positions of formation will be spread out around some preferred or average position. Where the formation of galaxies is such that the new galaxies are forming within the existing distribution of galaxies there will be an increase in matter density to the point where an event horizon forms.

The proposal is that the CMBR is coming from a distribution of galaxies located within the event horizon. The general appearance of the CMBR is consistent with a distribution of galaxies which explains the local variations which seem to have a scale of around one degree of arc. Features such as the "cold spot" are explained as due to the general absence of galaxies in that area. It is not unusual to find voids in galaxy distributions.

For galaxies close to the event horizon any radiation in the direction of the event horizon cannot cross the event horizon and would be reflected with the possibility of polarisation of the radiation.

We need to explain the precise cause of the frequency spectrum of the CMB which closely matches black body radiation. The galaxies exist in a particular region of space close to the event horizon and the effect of this on the radiation observed might be a factor to consider.

There are some unexplained anomalies in the CMBR data when considered against the Big Bang model explanation for the cause of the CMBR and these would also need to be explained in the context of the new hypothesis.

This is a work in progress which needs more general critique and analysis. However, the analysis that shows the existence of the event horizon in the position calculated is correct based on the Schwarzschild radius calculation and the matter density data.

The precise cause of the CMB itself is still uncertain but the location of the source close to the event horizon is confirmed.

Acknowledgment

The author would like to acknowledge the contribution from Ben Atkinson who pointed out the inconsistency in the first proposal for the source of the Cosmic Microwave Background Radiation.

References

1. Andrew Liddle (2003). An Introduction to Modern Cosmology (2nd ed.). John Wiley & Sons. p. 2. ISBN 978-0-470-84835-7.
2. William C. Keel (2007). The Road to Galaxy Formation (2nd ed.). Springer-Praxis. p.2. ISBN 978-3-540-72534-3
3. Alexander Friedmann (1923). Die Welt als Raum und Zeit (The World as Space and Time). Ostwalds Klassiker der exakten Wissenschaften. ISBN 978-3-8171-3287-4. OCLC 248202523..
4. Eduard Abramovich Tropp; Viktor Ya. Frenkel; Artur Davidovich Chernin (1993). Alexander A. Friedmann: The Man who Made the Universe Expand. Cambridge University Press. P. 219. ISBN 978-0-521-38470-4.
5. Lemaitre, Georges (1927). "Un univers homogène de masses constante et de rayon croissant rendant compte de la vitesse radiale des nébuleuses extra-galactiques". Annales de la Société Scientifique de Bruxelles. A47 (5): 49-56.
6. Helge Kragh: "The most philosophically of all the sciences": Karl

Popper and physical cosmology. Archived 2013-07-20 at the Wayback Machine (2012)

7. "Australian study backs major assumption of cosmology" 17 September 2012

8. "Simple but challenging: the Universe according to Planck". ESA Science & Technology. October 5, 2016 (March 21, 2013) Retrieved October 29, 2016.

9. Penzias, A. A.; Wilson, R. W. (1965). "A Measurement of Excess Antenna Temperature at 4080 Mc/s". The Astrophysical Journal. 142 (1): 419–421. Bibcode:1965ApJ...142..419P. doi:10.1086/148307.

10. Smoot Group (28 March 1996). "The Cosmic Microwave Background Radiation". Lawrence Berkeley Lab. Retrieved 2008-12-11.

11. Kaku, M. (2014). "First Second of the Big Bang". How the Universe Works. Discovery Science.

12. Wright, E.L. (2004). "Theoretical Overview of Cosmic Microwave Background Anisotropy". In W. L. Fredman. Measuring and Modelling the Universe. Carnegie Observatories Astrophysics Series. Cambridge University Press. P.291.

13. The Planck Collaboration (2014), "Planck 2013 results. XXVII. Doppler boosting of the CMB: Eppur is move", Astronomy, 571 (27): A27

14. Planck Collaboration (2018). Planck 2018 results. VI. Cosmological parameters.

15. "Planck reveals an almost perfect universe". Planck. Paris: ESA. 21 March 2013.

16. Kragh 1996, p. 319: "At the same time that observations tipped the balance definitely in favor of relativistic big-bang theory, ..."

17. Wright, Edward L. (24 May 2013). "Frequently Asked Questions in Cosmology: What is the evidence for the Big Bang?". Ned Wright's Cosmology Tutorial. Los Angeles: Division of Astronomy & Astrophysics, University of California, Los Angeles.

18. Landau & Lifshitz 1975, p. 228 "... the general theory of relativity... was established by Einstein, and represents probably the most beautiful of all existing physical theories."

19. O'Connor, J.J.; Robertson, E.F. (May 1996). "General relativity". History Topics: Mathematical Physics Index, Scotland: School of Mathematics and Statistics, University of St. Andrews.

20. Moshe Carmeli (2008). Relativity: Modern Large-Scale Structures of the Cosmos. Pp.92, 93. World Scientific Publishing

21. Grossmann for the mathematical part and Einstein for the physical part (1913). Entwurf einer verallgemeinerten Relativitatstheorie und einer Theorie der Gravitation (Outline of a Generalized Theory of Relativity and of a Theory of Gravitation), Zeitschrift fur Mathematik und Physik, 62, 225-261.

22. Biron, Lauren (7 April 2015). "Our universe is Flat". Symmetry magazine.org. FermiLab/SLAC

23. Kogut, Alan; et al. (December 10, 1993). "Dipole anisotropy in the COBE differential microwave radiometers first-year sky maps".

24. The Planck Collaboration (2020), "Planck 2018 results. I. Overview, and the cosmological legacy of Planck", Astronomy and Astrophysics, 641: A1

25. Hotokezaka, K.; et al. (8 July 2019). "A Hubble constant measurement from superluminal motion of the jet in GW170817".

26. Mukherjee, S.; Ghosh, A.; Graham, M.J.; Karathanasis, C.; et al. (29 September 2020). "First measurement of the Hubble parameter from bright binary black hole GW190521".

27. Pesce, D.W.; Braatz, J.A.; Reid, M.J.; Riess, A. G.; et al. (26 February 2020). "The Megamaser Cosmology project. XIII. Combined Hubble Constant Constraints". The Astrophysical Journal 891: L1.

28. Shajib, A. J.; Birrer, S.; Treu, T.; Agnello, A.; et al. (14 October 2019). "STRIDES: A 3.9 per cent measurement of the Hubble constant from the strongly lensed system DES J0408-5354".

29. Dutta, Koushik; Roy, Anirban; Ruchika, Richika; Sen, Anjan A.; Sheikh-Jabbari, M. M. (20 August 2019). "Cosmology with Low-Redshift Observations: No Signal For New Physics". Phys. Rev. D. 100 (10): 10351.

30. Overbye, Dennis (20 February 2017). "Cosmos Controversy: The Universe Is Expanding, but How Fast?". New York Times. Retrieved 21 February 2017.

31. Nussbaumer, H.; Bieri, L. (2011). "Who discovered the expanding universe?". The Observatory. 131 (6): 394-398.

32. Livio, M.; Riess, A. (2013). "Measuring the Hubble constant". Physics Today. 66 (10): 41

33. "Dark Matter". CERN Physics. 20 January 2012.

34. Trimble, V. (1987). "Existence and nature of dark matter in the universe". Annual Review of Astronomy and Astrophysics. 25: 425-472.

35. Bertone, G.; Hooper, D.; Silk, J. (2005). "Particle dark matter: Evidence, candidates and constraints". Physics Reports. 405 (5-6): 279-390

36. De Swart, J.G.; Bertone, G.; van Dongen, J. (2017). "How dark matter came to matter". Nature Astronomy. 1 (59): 0059.

37. Freeman, K.C. (June 1970). "On the Disks of Spiral and S0 Galaxies". The Astrophysical Journal. 160: 811-830

38. Zwicky, F. (1937). "On the Masses of Nebulae and of Clusters of Nebulae". The Astrophysical Journal. 86: 217-246.

39. Peebles, P. J. E.; Ratra, Bharat (2003). "The cosmological constant and dark energy". Reviews of Modern Physics. 75 (2): 559–

606. arXiv:astro-ph/ 0207347. Bibcode:2003RvMP...75..559P. doi:10.1103/ RevModPhys.75.559. in Conf. Rec. 1995 World Academy of Science, Engineering and Technology Int. Conf. Communications, pp. 3–8.

40. Paul J. Steinhardt; Neil Turok (2006). "Why the cosmological constant is small and positive". Science. 312 (5777): 1180-1183.

41. Riess, Adam G.; Filippenko; Challis; Clocchiatti; Diercks; Garnavich; Gilliland; Hogan; Jha; Kirshner; Leibundgut; Phillips; Reiss; Schmidt; Schommer; Smith; Spyromilio; Stubbs; Suntzeff; Tonry (1998). "Observational evidence from supernovae for an accelerating universe and a cosmological constant". Astronomical Journal. 116 (3): 1009–1038.

42. Perlmutter, S.;Aldering; Goldhaber; Knop; Nugent; Castro; Deustua; Fabbro; Goobar; Groom; Hook; Kim; Kim; Lee; Nunes; Pain; Pennypacker; Quimby; Lidman; Ellis; Irwin; McMahon; Ruiz-Lapuente; Walton; Schaefer; Boyle; Filippenko; Matheson; Fruchter; et al. (1999). "Measurements of Omega and Lambda from 42 high redshift supernovae". Astrophysical Journal. 517 (2): 565-586

43. K. Schwarzschild, "Über das Gravitationsfeld eines Massenpunktes nach der Einsteinschen Theorie", Sitzungsberichte der Deutschen Akademie der Wissenschaften zu Berlin, Klasse fur Mathematik, Physik, und Technik (1916) pp 189.

44. Richard Lewis. The Unification of Physics. International Journal of Recent Advances in Physics (IJRAP) Vol 9,No 4, November 2020

45. Barish, Barry C.; Weiss, Rainer (October 1999). "LIGO and the Detection of Gravitational Waves". Physics Today. 52 (10): 44

46. Castelvecchi, Davide (15 September 2015), "Hunt for gravitational waves to resume after massive upgrade: LIGO experiment now has better chance of detecting ripples in space-time", Nature, 525 (7569): 301-302.

5 THE EVOLUTION OF THE UNIVERSE - PREPRINT

Abstract

The following assumptions of the Big Bang theory are challenged and found to be false: the cosmological principle, the assumption that all matter formed at the same time and the assumption regarding the cause of the cosmic microwave background radiation (CMBR).

The evolution of the universe is described based on the conclusion that the universe is finite with a space boundary. The CMBR is coming from a source located within the event horizon of the universe. There is sufficient mass in the universe to create an event horizon at the Schwarzschild radius.

Galaxies form over time due to the energy released by the expansion of space. When a galaxy forms the matter formation is in the form of neutrons. Pairs of neutrons bond to form dark matter (dineutrons) and single neutrons decay to form protons and electrons which combine to form hydrogen.

Conservation of energy must consider total energy which is mass (+ve) plus energy (+ve) plus spacetime curvature (-ve) so that the total energy of the universe is always zero. The predominant position of galaxy formation moves over time from the centre of the universe towards the boundary so that today the majority of new galaxy formation is taking place beyond our horizon of observation at 14 billion light years.

The distribution of galaxies in a finite universe is such that there is a variation in gravitational acceleration depending on the distance from the centre of the universe. This gravitational acceleration results in the difference in recession velocity previously attributed to dark energy.

The Space Boundary Theory

The approach taken is to set aside the Big Bang theory and try to deduce the evolution of the universe from different starting assumptions. The Big Bang theory deduces the evolution of the universe by assuming a uniform progress back in time with a corresponding increase in temperature and density.

However, evolution moves forward in time and it is entirely possible that a different evolution path leads to the universe that we observe today. The starting configuration of the universe for the Space Boundary theory is a spherical region of space with a space boundary which then expands at a fixed rate which is the same rate as is observed today.

This puts the age of the universe at more than 300 billion years so we have to find a different explanation for the cosmic microwave background radiation (CMBR).

The Big Bang theory describes the CMBR as radiation arising from the Big Bang at a time 13.8 billion years ago. The Space Boundary theory describes the CMBR as radiation coming from a source located within the event horizon of the universe. This removes the time constraint on the age of the universe and allows the assumption of uniform expansion to be possible.

This requires a discussion about the event horizon of the universe. The idea of an event horizon arises directly from the equations of the General Theory of Relativity. Where there is a region of space with sufficient mass lying entirely within that region, spacetime is curved to such an extent that nothing can escape. Any high energy particles or radiation generated within this region of space will not be able to escape through the event horizon.

The condition that we observe radiation coming from the vicinity of the event horizon with a travel time of 13.8 billion years is that the radiation must have originated from a distance of 8.77 billion light years. Using this Schwarzschild radius it is possible to calculate the amount of mass in the form of galaxies that would be required to cause this event horizon.

Then taking that mass density forward in time to the present, it gives a matter density consistent with current observational data. So the universe that we observe is consistent with the existence of an event horizon causing the Cosmic Microwave Background Radiation.

The universe is finite with a boundary

This conclusion is reached by ruling out the possibility of an infinite universe or a universe which is finite with no boundary. An infinite universe would require infinite energy for matter formation and would violate the law of conservation of energy. A finite universe with no boundary is ruled out by the observation that the radiation from the CMBR follows a straight line path meaning that on the large scale of the universe the curvature is flat in the radial direction to the CMBR.

Given that the universe is finite with a boundary and making the symmetrical assumption that the boundary is spherical this means that the universe has a centre. The space within the boundary of the universe is a spherical region of space which is expanding away from the centre of the universe and the recession velocity of each point of space is proportional to the distance from the centre of the universe.

Our position in the universe

The CMBR provides a useful frame of reference for the universe called the CMB rest frame. This means that there will be a point located at the centre of the universe which is at rest with reference to the CMB rest frame.

When the CMBR is measured in great detail it reveals that the Milky Way galaxy is moving at a speed of 552 km/sec with reference to the CMB rest frame.

The universe is expanding uniformly in all directions at a steady rate of 1 part in 14 billion light years per year per light year. Each galaxy in the universe can be thought of as being embedded in this expanding frame of reference where the recession velocity of each galaxy depends on the distance from the centre of the universe.

Using the velocity of the Milky Way galaxy and the rate of expansion of space we can locate the centre of the universe at a distance of approximately 26 million light years from the Milky Way galaxy. Put another way, this means that the Milky Way galaxy is located 26 million light years from the centre of the universe and is moving away from the centre of the universe at 552 km/sec.

Galaxy formation

Going back 200 billion years, there were no galaxies, just an expanding spherical region of space. Galaxy formation started an estimated 126 billion years ago. The source of energy for the formation of galaxies is the energy released from the expansion of space. The galaxy formation events start closest to the centre of the universe with the first galaxy forming 126 billion years ago. The predominant position of formation of new galaxies moved away from the centre of the universe towards the boundary so that by now the majority of galaxy formation takes place beyond the observation horizon at 14 billion light years.

A galaxy formation event creates a spherical region in which the release of energy from space leads to the formation of neutrons some of which immediately bond to form di-neutrons which consist of two neutrons bonded together. These di-neutrons are what we observe as dark matter. Single neutrons will decay over a short time period to create a proton an electron and a neutrino. The protons and electrons can together form hydrogen atoms and then hydrogen molecules. Some of the material (hydrogen and dark matter) will fall under gravity towards the centre of the gas cloud and result in the formation of a large neutron star which will be viewed externally as a super massive black hole.

Dark Energy

The existence of dark energy has been proposed to explain the variation in the recession velocity of more distant galaxies. The recession velocity of galaxies was expected to follow Hubble's law where the recession velocity of a distant object is proportional to its distance.

An accurate measurement of distance was made possible by observing distant supernovae to measure the luminosity of the supernova event. It was found that the relationship between recession velocity and distance was not proportional with a different value of the Hubble constant coming from the observation of distant galaxies.

In the Big Bang model the universe is considered to be isotropic so the expansion of space must be the same everywhere. The only way of accounting for this discrepancy is to assume that the rate of expansion has been slower in the past and this is explained as caused by Dark Energy.

Moving to a model of the universe in which space is finite with a boundary, this defines a unique frame of reference and the universe has a centre. Then it is possible to consider variation in the Hubble constant value with distance from the centre of the universe.

The conclusion is that there is no dark energy and the expansion of space is constant over time and distance. The variation in the observed value of the Hubble constant with distance is due to the movement of galaxies through space relative to the expanding CMB rest frame.

Conclusions

The evolution of the universe has been described based on the conclusion that the universe is finite with a space boundary. This conclusion is reached by ruling out the possibility of an infinite universe or a universe which is finite with no boundary. In a finite universe the centre of the universe can be located with reference to the Milky Way galaxy using the speed relative to the Cosmic Microwave Background Radiation (CMBR) rest frame and Hubble's law. This places the Milky Way galaxy at a distance of approximately 26 million light years from the centre of the universe.

There was no Big Bang and the age of the universe is greater than 300 billion years. Galaxies then formed over time due to the energy released by the expansion of space throughout the universe. The nature of Dark Matter is explained as the dineutron. This is two neutrons bonded together.

There is no dark energy and the expansion of the universe is not accelerating.

List of Appendices

1. Count of the number of galaxies with increasing distance

2. The variation of the radius of the universe with time

3. Formation of mass in the universe

4. Observed matter density

5. The age of the universe

6. Dark Matter analysis

7. Black Holes

8. Cosmic Microwave Background Radiation

9. The Cosmological Principle

10. The formation of the solar system

11. Sequential derivation of the Space Boundary Theory

12. Event Horizon and Observation Horizon

13. Cosmic Rays

14. Galaxy formation principles

15. Galaxy formation analysis

16. Event horizon formation analysis

17. Galaxy recession velocity and Dark Energy

18. Galaxy distribution

19. Elliptical and Spiral Galaxies

20. Our position in the universe

21. Galaxy distribution near the centre of the universe

22. Schwarzschild Radius and Matter Density

23. Expansion Factor K

Appendix 1 Count of the number of galaxies with increasing distance

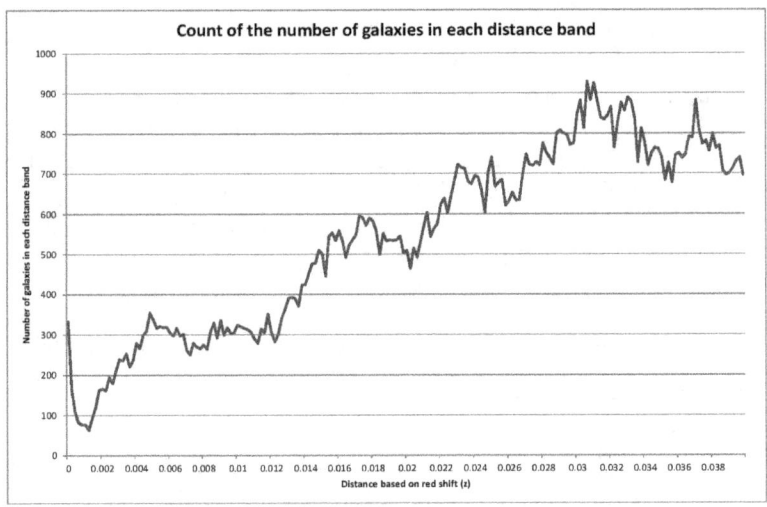

The plots above show the count of the number of galaxies in distance bands of approximately 2.75 million light years (red shift steps of $z = 0.0002$). The counts are calculated based on data from the NASA Extragalactic Database (NED). The range of the data is out to $z = 0.04$ which is approximately 550 million light years.

The NED data was accessed choosing to exclude objects with the prefix "2dFGRS". These objects are excluded because they are part of a very detailed study of a small region of the sky. The graph does not include error bars because the graph plots are the exact galaxy counts for each distance band based on the NED data. This data was first obtained from NASA in 2007 and subsequently accessed again in 2017 with no noticeable change in the results.

As of 2018 additional measures are needed to reproduce these results based on the NASA Extragalactic Database data. Firstly it is necessary to exclude the point source data introduced by 2MASS as well as the 2dFGRS data points mentioned above. Also the items which are flagged with the redshift flags PHOT (estimated using photometry) and SPEC (an explicitly declared spectroscopic value) are to be removed from the galaxy counts. The data can be downloaded as a complete set with no exclusions and the unwanted data points filtered out once all the data is captured in a spreadsheet.

After filtering out these data points, the galaxy counts by distance revert to the pre 2018 levels. These galaxy counts are considered to be more representative of the mass distribution of galaxies by distance and do not include estimated red shift values. The count of galaxies by distance band can then be obtained as above. The observational data is consistent with a process of galaxy formation in which the position of formation of galaxies moves from the centre of the universe towards the boundary.

Galaxies adopt the recession velocity appropriate to their distance from the centre at the time of formation since they form in the expanding frame of reference of the CMB rest frame. The number of galaxies in each distance band does not follow a square law as would be expected with a uniform distribution of galaxies. The volume of each distance band is a spherical shell which is proportional to the square of the distance.

Appendix 2 The variation of the radius of the universe with time

The proposal analysed here is that the recession velocity of the boundary from the centre of the universe is proportional to the radius. This means that to a first level of approximation, the rate of expansion of space is constant over time and distance.

Then $\Delta R / \Delta T = K R$ where K is a positive constant,

This means that $\Delta T = (1/ K) \Delta R / R$

Integrating this expression we get: $T = (1/ K) \log (R)$ plus a constant.

Since the time value can be set to an arbitrary reference point we take the value of $T = 0$ to be when the radius of the universe was 1 light year. This makes the constant of integration zero.

From the above equation for T we get: $R = \exp (K T)$

Also differentiating to get the recession velocity we get: $dR/dT = K \exp (K T)$

To obtain the value of K we use the observed local expansion of space as a first approximation. We also assume as a first approximation that the expansion of space is uniform out to the boundary. Currently the observation of galaxy recession velocities shows that $(dR/dT) / R$ is equal to 1/(14 billion) which is therefore the value of K.

(Note that the value of 1/(14 billion) light years per year per light year is taken from observations of the value of the Hubble constant. This value of K corresponds to a Hubble constant value of 69.84 km/s per Mpc. If subsequent observations suggest a different value for K this will affect the numerical results of this paper by a few percentage points.)

So we have two equations relating T and R.

$T = (1/ K) \log (R)$

R = exp (K T)

To get an initial estimate of the current value of T and R, the observational evidence of the Cosmic Microwave Background Radiation (CMBR) is used.

The CMBR is coming from the event horizon of the universe and we observe the CMBR having travelled for 13.8 billion years at the speed of light.

The position of the event horizon 13.8 billion years ago can be calculated by considering the expansion of space at one part in 14 billion light years per year per light year during the travel time.

The calculation divides the total travel time of the CMB radiation into equal intervals dt in such a way that the contribution to the distance measurement at time t is given by dt / exp (K t) for each time segment dt. Then by integrating dt / exp (K t) from t=0 to t=13.8 billion we obtain the result 14*(1 - exp (-13.8/14)) billion light years for the position of the event horizon at a time 13.8 billion years ago.

The event horizon must have been at approximately 8.775584 billion light years from the centre of the universe at the time 13.8 billion years ago. The CMB radiation has covered the original separating distance of 8.775584 billion light years during a travel time of 13.8 billion years because over that time interval the expansion of space has increased the distance to be covered.

For the CMB to be visible, the radius to the boundary must have been greater than the radius to the event horizon at a time 13.8 billion years ago.

The minimum value for R at a time 13.8 billion years ago is therefore 8.77 billion light years.

Using T = (1/ K) log (R)

T is currently greater than 320.53 + 13.8 = 334.33 billion years.

Using R = exp (K T)

The radius of the universe (R) is currently greater than 23.52 billion light years.

Appendix 3 Formation of Mass in the Universe

As the universe expands, the total energy (mass plus energy plus spacetime curvature) must remain the same. The expansion of space results in an increase in the radius of curvature of space at every point in space.

There will be a relationship between the total mass formed within the universe and the volume of the universe. The proposed relationship is that the total mass of galaxies in the universe is proportional to the volume of the universe.

The formula M proportional to R^3 together with $R = \exp(K\,T)$ implies that every 14 billion years the mass of all the galaxies in the universe increases by a factor of $e^3 = 20.0855$. This new galaxy formation is currently taking place mostly beyond 14 billion light years so beyond our observation horizon.

Appendix 4 Observed Matter Density

We can look at the data presented in Appendix 1 and count the number of galaxies out to 550 million light years. The data from the NASA Extragalactic database indicates that there are 110,000 galaxies in this spherical region of radius 550 million light years.

The matter density in this region can be calculated:

The estimated mass of the Milky Way galaxy is 0.8 to 1.5 x 10^{12} solar masses

One solar mass is 1.988 x 10^{30} kg

So the estimated mass of the Milky Way galaxy is 1.6 to 3 x 10^{42} kg

The Milky Way galaxy is taken as representative of a typical galaxy mass which is taken to be 2 x 10^{42} kg.

The total mass of the 110,000 galaxies counted in Appendix 1 is estimated to be approximately 2.2 x 10^{47} kg

The volume of this region is $4/3\,\pi\,R^3$ where R = 550 million light years.

The volume is 0.59 x 10^{75} m^3.

The matter density is 0.373 x 10^{-27} kg m^{-3}

For there to be an event horizon at 8.775584 billion light years at a time 13.8 billion years ago there must have been a mass (using $R = 2Gm/c^2$) of 5.59 x 10^{52} kg present. This corresponds to an average matter density of 23.32 x 10^{-27} kg m^{-3} at a time 13.8 billion years ago.

The uniform expansion of the universe over the last 13.8 billion years has been $\exp(13.8/14) = 2.6797$ which corresponds to a volumetric expansion by a factor of 19.2429.

The expected observed average matter density now (23.32/19.2429) should be 1.2119 x 10^{-27} kg m-3 based on the presence of the event horizon. This is more likely to be the accurate estimate for matter density since the observational data of Appendix 1 only goes out to 550 million light years.

This shows that the conclusion that there was sufficient mass in the universe 13.8 billion years ago to create an event horizon is consistent with current observational data for matter density.

Appendix 5 The Age of the Universe

The formulas for the expansion of space developed in Appendix 2 are:

$T = (1/K) \log (R)$

$R = \exp(K\ T)$

Where $K = 1/14$billion Light Years / Year per Light Year (LY/Y per LY)

In an elapsed time of 14 billion years the expansion of any region of space more than doubles:

Suppose $T_2 = T_1 + 14$ billion years.

Then $\log (R_2) = \log(R_1) + 1$

$R_2 = R_1\ x\ e$ where e is the exponential constant and has a value of approximately 2.71828.

So if we were to assume that the equations of expansion operated down to the smallest scales then if we put $R_1 = 1$cm then $R_2 = 2.71828$ cm.

This implies that it takes 14 billion years for the universe to expand from 1 cm to just under 3 cm. It is not safe to assume that the expansion of space at such small scales follows the uniform expansion equations.

The approach taken is to apply the expansion equation for positive values of T only, meaning that the equation $R = \exp(K\ T)$ is only used for values of R greater than 1 light year. The characteristics of the expansion of the universe for value of R of less than 1 light year require further analysis outside the scope of this paper.

The analysis of the evolution of the universe in the Space Boundary Theory then starts when $T = 0$ and $R = 1$ light year. The universe is at least 334 billion years old as calculated in Appendix 2.

The question of whether the age of the universe is finite or infinite is outside of the scope of the Space Boundary Theory. The universe may or may not have had a beginning.

Appendix 6 Dark Matter analysis

During galaxy formation the initial galaxy formation event results in the formation of neutrons in numbers corresponding to the total mass of the galaxy. We know that a single neutron will decay into a proton and an electron after an average period of around 15 minutes.

What would happen if two neutrons collided before the decay into a proton and an electron? Under the right conditions two neutrons will bond to form a di-neutron. We also know that neutrons involved in atomic bonds do not decay so easily into a proton and electron.

So the formation of di-neutrons during galaxy formation would have two effects. Firstly we would expect these di-neutrons to fall under gravity and form the neutron star within the central black hole. Secondly, di-neutrons would not be detectable by photons and would pervade the galaxy halo thus being an ideal candidate for dark matter.

Any material that falls under gravity towards the centre of the galaxy will increase the mass of the super massive neutron star which we observe as a super massive black hole at the centre of each galaxy.

The idea of di-neutrons as dark matter depends on the binding energy of two neutrons being positive. The binding energy corresponds to the mass defect associated with the neutron to neutron bond.

There may be other factors to take into account when considering the binding of neutrons. Considering the neutron as a looped wave in spacetime with three wavelengths in the loop, it may be necessary for the spin axes of the neutrons to be aligned for them to bond. This alignment could well be the case if the neutrons are created from a helical wave disturbance of spacetime during the galaxy formation event. The helix structure would decay into individual neutrons with the spin axes aligned.

To understand the proportion of dark matter compared with hydrogen, consider the formation of a linear row of single neutrons:

NNNNNNNNNNNNNNNNNNNNNNNNNNNNNNNNNNNNNNNNNN
NNNNNNNNNNNN

These will try to pair (PP) with an adjacent neutron but only if that neutron is not already paired. This is the first pass of pairing with variable numbers of adjacent neutrons remaining unpaired:

NPPNPPNPPNNNNNNPPNNPPNNNPPNPPNNPPNNNNPPPPNN
PPPPNNN

Any single N will remain as a single N, a double N will pair (PP) and three Ns together will result in (NPP), four Ns together might result in NPPN or PPPP, 5 Ns together will likely result in PPNPP and so on.

So after further pairing of neutrons we would expect:

NPPNPPNPPPPNPPNPPPPPPPPPNPPNPPPPPPPNPPNPPPPPPPPPPPN
PP

The single neutrons will decay to a proton and electron and the paired neutrons are the dark matter particles which represent around 80% of the mass (40 out of 50 in this example).

Appendix 7 Black Holes

The description of the formation of galaxies includes the formation of the super massive black hole at the centre of every galaxy. The formation of the galaxy starts with the formation of a number of neutrons equivalent to the total mass of the galaxy. A proportion of the neutrons form di-neutrons and move under the effect of gravity to form the central black hole.

In this section we investigate the hypothesis that all black holes contain neutron stars and that the event horizon of the black hole is caused by the mass and high density of the neutron star within the event horizon.

One prediction from this hypothesis is that we should be able to see a range of sizes of neutron stars and black holes but any black hole will always be greater in mass than the largest observed neutron star. At present the largest observed neutron star is 1.97 solar masses and the smallest observed black hole is 3.8 solar masses.

We can estimate the size of neutron star where the event horizon is exactly at the surface of the star. The following symbols are used in the calculation:

N is the number of neutrons in the neutron star

m is the mass of a neutron 1.675×10^{-27} kg

r is the radius of a neutron

R is the Schwarzschild radius of the star.

M is the mass of the star

Schwarzschild radius formula gives: $R = 2GM/c^2$

We can start with $M = Nm$ for the mass of the star

$R = 2GNm/c^2$ gives the radius of the star

The volume of the star is $4/3 \pi R^3$

The volume of the star is also $N \times$ volume of the neutron $= 4/3 \pi r^3$

So $4/3 \pi R^3 = N \times 4/3 \pi r^3$ Therefore $R^3 = N r^3$

$(2GNm)^3 = c^6 N r^3$

$N^2 = c^6 r^3 /(2Gm)^3$

Now we have to decide on the value of r to use. The effective radius value calculated in Appendix 4 of the Unification of Physics is 0.630058×10^{-15} m. This gives a value of N as approximately 4.03×10^{57}

This number of neutrons has a mass of 6.751×10^{30} kg which is 3.4 solar masses. The value of R is given by $N^{1/3}$ r which is 10.027 kilometres. The minimum radius for a blackhole is approximately 10km.

This lends support to the theory that all black holes contain neutron stars which cause the event horizon of the black hole. The calculated value for the largest observable neutron star (3.4 solar masses) and the smallest possible black hole (3.4 solar masses) is consistent with observation.

For black holes larger than 3.4 solar masses, the position of the event horizon at the Schwarzschild radius increases in proportion to the mass whereas the radius of the neutron star within the event horizon increases in proportion to the cube root of the mass. So for a supermassive black hole of 3.4 million solar masses the radius of the neutron star would be approximately one thousand kilometres whereas the Schwarzschild radius would be 10 million kilometres.

In summary, the space boundary theory and the spacetime wave theory point to the conclusion that the internal structure of a black hole is not a singularity but a neutron star of a mass corresponding to the observed mass of the black hole. There has been considerable success with gravitational wave detectors which detect merging black holes and merging neutron stars. The neutron stars observed are always less than 3.4 solar masses and the merging black holes are all greater than 3.4 solar masses. The production of gravitational waves is the same in the two cases because, in both cases, we are observing merging neutron stars.

When two black holes merge the angular momentum of the system causes a rotating black hole to form. This is in fact a merger of two neutron stars which results in a rotating neutron star inside the event horizon. Where the mass inside the event horizon is rotating the calculation of the event horizon at the Schwarzschild radius has to be modified to take into account the rotation of the mass.

The event horizon may not be closed at the axis of rotation and the jets which are characteristic of rotating neutron stars can also be visible coming from the axis of the merged black hole. This opening of the event horizon at the axis also applies to the merging of galaxies where the central black holes combine and under the right conditions of mass and rotation the jets can emerge from the central black hole of the merged galaxy along the axis of rotation.

The neutron star may also contain a proportion of protons. The proton is approximately the same mass and volume as the neutron so the above calculation for the event horizon still applies. The electrostatic repulsion of the positively charged protons is overcome by the gravitational forces of the star. In the case of a rotating neutron star with a proportion of protons, the effect of the protons is to create an electric charge moving in a circular path, which then creates a strong magnetic field.

Neutron stars vary in their total mass, their speed of rotation and the proportion of protons and this variation then affects the ability of the neutron star to generate high energy jets emerging at the two poles along the axis of rotation.

Appendix 8 Cosmic Microwave Background Radiation

The Cosmic Microwave Background Radiation (CMBR) has been mapped in detail and has been used as the key piece of evidence in support of the Big Bang theory. In the Big Bang theory the CMBR is believed to be the radiation coming from the Big Bang itself following a process called recombination.

In the Space Boundary theory of the evolution of the universe it is proposed that the CMBR is radiation coming from the vicinity of the event horizon of the universe. The General Theory of Relativity (GR) provides equations which allow us to calculate the curvature of spacetime due to a given distribution of mass.

These equations have a solution first proposed by Schwarzschild: $R = 2Gm/c^2$

This equation can be used to calculate the event horizon for a given distribution of matter. The event horizon is at the radius R for a mass m. Nothing, not even light can escape beyond the event horizon. This formula is used to calculate the event horizon for a black hole and in this case we are looking from the outside towards a black hole which we can detect because of its gravitational effect on other objects.

The equation can also be applied to the distribution of mass within the universe where we are observing from within the event horizon and in this case it shows that there will be a radius from which matter and radiation cannot escape.

Based on the observation that the radiation was emitted 13.8 billion years ago we can calculate that at the time of the emission of this radiation the distance to the source was 8.77 billion light years. This implies an event horizon at a radius of 8.77 billion light years at a time 13.8 billion years ago.

We can then calculate the matter density required to create an event horizon at this distance. From this we can project that matter density forward in time assuming an expansion of 1/(14 billion) light years per year per light year and we find that the matter density observed today should be around 0.724 hydrogen atoms per cubic metre.

So the analysis is consistent with the CMB radiation coming from a source in the vicinity of the event horizon. What is the nature of this source of radiation? This is still a subject of investigation but the following hypothesis is presented.

The formation of galaxies takes place at points in space which are located progressively further from the centre of the universe. The model suggests that the number of galaxies increases by a factor of 20 every 14 billion years and the average position of formation of these galaxies moves from the centre of the universe outwards towards the boundary.

The actual positions of formation will be spread out around some preferred or average position. Where the formation of galaxies is such that the new galaxies are forming within the existing distribution of galaxies there will be an increase in matter density to the point where an event horizon forms.

The proposal is that the CBMR is coming from a distribution of galaxies located within the event horizon. The general appearance of the CMBR is consistent with a distribution of galaxies which explains the local variations which seem to have a scale of around one degree of arc. Features such as the "cold spot" are explained as due to the general absence of galaxies in that area. It is not unusual to find voids in galaxy distributions.

For galaxies close to the event horizon any radiation in the direction of the event horizon cannot cross the event horizon and would be reflected with the possibility of polarisation of the radiation.

We need to explain the precise cause of the frequency spectrum of the CMB which closely matches black body radiation. The galaxies exist in a particular region of space close to the event horizon and the effect of this on the radiation observed might be a factor to consider.

There are some unexplained anomalies in the CMBR data when considered against the Big Bang model explanation for the cause of the CMBR and these would also need to be explained in the context of the new hypothesis.

This is a work in progress which needs more general critique and analysis. However, the analysis that shows the existence of the event horizon in the position calculated is correct based on the Schwarzschild radius calculation and the matter density data. The precise cause of the CMB itself is still uncertain but the location of the source close to the event horizon is confirmed.

The Big Bang theory stands or falls depending on the explanation for the CMBR. In the Big Bang theory the CMBR is electromagnetic radiation as a remnant from the early stage of the universe dating from the epoch of recombination 370,000 years after the Big Bang at a redshift of $z = 1100$.

In Big Bang cosmology, recombination refers to the epoch at which charged electrons and protons first became bound to form electrically neutral hydrogen atoms. Prior to this they were in the form of a plasma which was effectively opaque to electromagnetic radiation due to Thomson scattering. After recombination photon decoupling occurs and photons may freely travel through the universe and according to the Big Bang theory, this is what we observe as the CMBR.

The temperature at which recombination is supposed to have occurred is taken not from any physics experiment but from the calculated temperature which would apply to a red shift of $z = 1100$ when the known temperature of the CMBR today is 2.725 degrees Kelvin.

The apparent look back time for the CMBR of 13.8 billion years is explained as follows: "the Big Bang happened everywhere and we observe the photon decoupling from the region of space at a corresponding distance". But this implies that the CMBR radiation started and then stopped which implies a flash of radiation as the temperature drops through 3000 degrees Kelvin.

If this were the case, we should be able to observe this physical phenomenon of a flash of radiation at 3000K in the laboratory today by observing a plasma within an experimental hydrogen fusion reactor. I have heard of no such observation and without this the Big Bang theory falls.

Appendix 9 The Cosmological Principle

In modern physical cosmology the cosmological principle is the notion that the spatial distribution of matter is homogeneous and isotropic.

Homogeneous: The property wherein no location can be distinguished from any other.

Isotropic: The property wherein no direction can be distinguished from any other; also the condition of being perfectly uniform.

There are some observations which directly contradict this principle. Firstly the existence of the CMB rest frame and the ability to locate a CMB rest point at the centre of the universe is significant. The centre of the universe is a particular point in space and this is the point of rest such that the expansion of the universe is with reference to this centre.

It is significant that there are no galaxies within a distance of 9.2 million light years from the centre. The galaxy distribution pattern of Appendix 21 shows the special nature of the centre of the universe in terms of galaxy distribution.

The galaxy distribution pattern of Appendix 1 shows a galaxy distribution pattern of increasing numbers of galaxies centred on the centre of the universe.

The existence of a space boundary and an event horizon also means the universe is not homogeneous and isotropic.

The cosmological principle is not correct.

This is an important conclusion because the Friedmann equations depend on the assumption that the universe is homogeneous and isotropic.

Appendix 10 The Formation of the Solar System

We would like to understand how exceptional are the coincidences which resulted in the formation of the Earth with its ability to support life. This description is not dependent on a general theory of the evolution of the universe since all theories propose star formation from clouds of gas and dust.

This proposal rejects the theory that the planets accumulated their mass from smaller objects in a cloud of gas and dust coming together under gravity. This is not possible because the trajectory of material within a cloud of gas and dust would be to fall towards and into the central star.

By considering the angular momentum of the sun, it is possible to project back in time to calculate the angular rotation of the original cloud of gas and dust using the conservation of angular momentum. The calculation shows that the rotation of the original gas cloud had a very small angular velocity. Then if you consider the trajectory of a particle within the gas cloud, it will accelerate under gravity towards the centre of mass of the cloud. The acceleration creates a radial component of velocity towards the sun which is very much greater than the rotational component with the result that every particle in the gas cloud must end up in the sun with no possibility of the direct formation of a disk in circular orbit.

If planets formed from the accretion of smaller objects, we would also need to explain how this created a molten core. Instead it is proposed that the iron core of the Earth came directly from material in the core of the sun formed from the cloud.

The planets must have formed from material which came originally from the cloud of gas and dust but only after it had combined into one or more stars. The heavier elements in the solar system must already have been present in the cloud of gas and dust and the cloud of gas and dust would itself have been formed from an exploding star or several generations of stars prior to the formation of the sun 4.6 billion years ago.

One important observation of star formation in general is that it is not uncommon to find binary and multiple star systems. 'It is estimated that approximately one third of the star systems in the Milky Way galaxy are binary or multiple with the remaining two thirds being single stars'. (source Wikipedia). The proposal here is that the solar system is a special case where the angular momentum is too small to form a binary system or multiple star system but too large to form a single stable star without planets.

The first step in star formation is for the gas and dust cloud to come together and the temperature to rise sufficiently for nuclear reactions to start. The cloud of gas and dust continues to fall under gravity into the sun adding to its speed of rotation and creating a sphere of empty space surrounding the sun as the gas cloud collapses.

The heavier elements will tend to fall under the gravitational forces of the sun to form the core of the sun. As the heavier material falls towards the centre of the sun it conserves its angular momentum and so the core of the sun will be spinning faster than the surface of the sun.

The core then would adopt a flattened spherical shape and given a sufficient speed of rotation a large body of material would be projected away from the sun, starting with the outer part of the spinning core in the plane at right angles to the axis of spin. This takes with it a body of material including all the elements present in the sun from the core to the surface.

This is effectively a large mass ejection from the sun which could be somewhere between 1% and 10% of the total mass of the sun. Given that the sun prior to this mass ejection contained separate layers of elements organised by atomic weight, we could imagine what would happen when the oxygen layer came into contact with the hydrogen layer.

This body of material which is spun away from the sun, leads to the formation of all the planets and asteroids. We need to understand why the planets' orbits are approximately circular and not elliptical as one might expect for material following a path starting from the sun.

It is proposed that there was an amount of mass much greater than the total mass of the planets spun out from the sun. The mass of the sun is 1047 times greater than the mass of Jupiter so we could imagine a body of material somewhere between 10 and 100 times the mass of Jupiter thrown out from the sun due to the fast rotation of the core of the sun. If such a mass of rotating material separated from the sun then the planets would be left behind along its path. The material spun off from the sun will contain all of the elements existing within the sun.

The trajectory of the spun off material would have been as far as the Kuiper belt which is the extent of the material which orbits the sun in the same orbital plane. The spun off body of material would itself be rotating and would leave a portion of its material along its path away from the sun in the form of entire planets as well as asteroids. There would have been gravitational forces coming from the main body of material and from the sun pulling in opposite directions. The main body of this material is on an elliptical orbit starting from the sun and so we would expect it to fall back into the sun leaving the planetary material and asteroids in orbit around the sun. The planets and asteroids are all moving in a single plane and this plane is approximately at right angles to the axis of rotation of the sun.

It is likely that the atomic elements within the sun prior to this spin out were positioned in layers by atomic weight. Then as this body of spinning material was ejected from the sun, the oxygen layer was able to mix with the abundant hydrogen to form water.

The material for the planets came directly from a the sun with the core of the planets coming from the core. The planets and moons formed directly in their current size with only a small amount of subsequent accretion of material. The root cause of the formation of the planets was an excess of angular momentum which prevented the sun forming as a single star without planets.

A substantial part of the material of planetary formation formed the large outer planets of the solar system Jupiter, Saturn (gas giants) and Uranus and Neptune (ice giants). These planets have sufficient mass to hold hydrogen and helium under their gravitational force. These outer planets contain the heavier elements in their core and in the orbiting moons.

The material for the inner planets included hydrogen and helium and also heavier elements such as oxygen, nitrogen, iron, silicon and all the other naturally occurring elements on Earth. The material released for inner planet formation would have been in fluid form and would quickly adopt a spherical shape under its own gravity. During the initial formation of the inner planets Mercury, Venus, Earth and Mars there would be a considerable percentage of hydrogen and helium present. However, the mass of the inner planets would be insufficient to hold hydrogen and helium in the atmosphere.

At the same time as the planets were formed there would have been smaller clumps of material in the form of asteroids throughout the region from the sun to beyond the outer planets. This resulted in the asteroid belt between the inner and outer planets and the Kuiper belt beyond the outer planets. Any asteroids located in the orbital region of the inner and outer planets would be cleared by collision with the planets in that region. Objects orbiting the sun tend to be in approximately circular paths because any strongly elliptical path would lead the object to collide with a planet or fall into the sun.

The Earth is the most massive of the inner planets and has sufficient mass to hold nitrogen and oxygen in the atmosphere. At the time of the formation of the Earth the mass of material for planetary formation would be in fluid form (gaseous or liquid) and planets and moons formed at the same time. The water found on Earth came directly from the oxygen and hydrogen in the stellar material which formed the planets.

Given a correct theory for the formation of the solar system we can better estimate the likelihood of a similar system developing elsewhere. We would expect that it depends on the mass and element composition of the cloud of gas and dust and also on its precise rotation prior to the formation of the sun. We can consider the possible history of generations of star formation that resulted in the combination of elements that were in the cloud of gas and dust which formed the solar system.

We can assume that a similar process of planetary formation would occur elsewhere, but the precise dynamics leading to the formation of particular sized planets at a particular distance from the star with a particular material composition is difficult to predict.

Appendix 11 Sequential Derivation of the Space Boundary Theory

The aim is to present a logical sequence of steps to arrive at the proposed description of the structure and evolution of the universe in this paper. By presenting the theory in a logical sequence of steps it provides the opportunity for others to challenge the Space Boundary Theory at each step.

For each step an assertion is made and then demonstrated by observational evidence together with the theory of gravity described by the General Theory of Relativity. The Theory also conforms to the law of conservation of total energy and the laws of thermodynamics.

Step 1: The universe must be finite with a space boundary.

To confirm this assertion we have only to rule out the other possibilities which are that the universe is infinite or that the universe is finite with no space boundary.

We can show that the universe cannot be infinite by considering the gravitational effect of an infinite distribution of matter with a matter density similar to the matter density that we observe. Mass has the effect of curving spacetime and this effect is cumulative. If we take a point in this infinite space and consider a spherical region of space with the chosen point as its centre, then for a sufficiently large radius the gravitational effect within the spherical region will cause an event horizon to form.

If such large radius event horizons were prevalent everywhere then this would be apparent in our observations. The possibility of event horizons located at every point in the universe presents a physical contradiction so the universe could not be infinite. Furthermore, an infinite distribution of matter in the universe represents a violation of the law of conservation of energy.

The universe does not have a form which is finite with no space boundary. This possibility is ruled out by the observation of the Cosmic Microwave Background Radiation (CMBR). Regardless of the explanation for the CMBR (see step 2), the observation of the CMBR shows that space is flat on a large scale between the source of the CMBR and our point of observation. If space is flat out to the CMBR then this rules out the possibility that the universe has a large scale curvature which would be required if there were no boundary.

So, the universe must be finite with a space boundary.

Step 2: The Cosmic Microwave Background Radiation (CMBR) is coming from within the event horizon of the universe

The assumption of the Big Bang theory of the evolution of the universe is that the CMBR arises from the Big Bang itself but the radiation only becomes visible when the temperature drops to a point where radiation is able to propagate (at recombination).

Staying with the conclusion that the universe must be finite with a boundary it is possible to calculate the change in the radius of the universe as time passes. The observation that the expansion of space is approximately uniform means that the radius of the universe can be estimated for earlier periods.

The Big Bang theory proposes that the CMBR was emitted simultaneously through space from every point (the Big Bang happened everywhere) and that the apparent distance of the source of the CMBR increases over time at one light year per year. The increase in the radius of a finite universe during the early period following recombination would be very much less than one light year per year so soon after recombination, the CMBR would no longer be visible anywhere as the source distance has passed beyond the finite boundary.

Instead, consider the finite distribution of matter within the boundary of a finite universe and apply the Schwarzschild Radius formula to locate the event horizon. The CMBR is coming from a source within this event horizon.

By changing our understanding of the source of the CMBR we are liberated from the Big Bang theory conclusion about the age of the universe. We can calculate the events leading to the formation of matter in the universe without constraints on the age of the universe.

Step 3: The formation of matter in the universe was not simultaneous with the formation of spacetime

By changing our understanding of the source of the CMBR we depart from the Big Bang theoretical understanding of the formation of matter. To conform to the law of conservation of energy it is necessary for a source of energy to develop before matter and radiation can exist.

The law of conservation of energy has to be revised to take into account the total energy which includes mass plus energy plus spacetime curvature. The total energy of the universe must always be zero to avoid violating the law of conservation of total energy. This means that the positive mass/energy of matter must always be balanced by the potential (negative) energy of spacetime curvature.

The expansion of space then becomes the source of energy for the formation of matter. Consider the region of space comprising the entire universe at a time before matter formation. The radius of the universe is steadily increasing which means that the radius of curvature of space everywhere in space is increasing.

This change in the energy of spacetime curvature must be balanced by the formation of matter in order for conservation of total energy to be maintained.

Step 4: Galaxies form at points in space where the tension in the fabric of space is at a maximum

The precise mechanism for matter formation is proposed based on evidence of the structure of galaxies which seem to form at points in space resulting in an expanding spherical region with a structure comprising a central black hole, a distribution of stars and a galaxy halo of dark matter (di-neutrons).

Appendix 12 Event Horizon and Observation Horizon

Throughout this paper the term event horizon refers to the horizon occurring at the Schwarzschild radius around a distribution of mass. It is at a specific location in space independent of any observer. Radiation emitted from outside the event horizon can pass through the event horizon towards the centre of the distribution of mass but no radiation can pass through the event horizon in the direction away from the centre of the distribution of mass.

The term observation horizon is taken to mean the horizon beyond which emitted radiation cannot be observed due to the expansion of space. If we take a point at a distance of 14 billion light years from our point of observation and light is emitted from this point, then in the first year the light will travel one light year. In that time the distance yet to travel to the point of observation will have increased by one light year so the radiation will never reach the point of observation. The observation horizon is at 14 billion light years.

Appendix 13 Cosmic Rays

Cosmic rays are highly energetic particles which are detected through their collisions in the upper atmosphere of the Earth.

To date there has been no explanation for the very high energies that some of these particles exhibit. Previously there were no known cosmic events which could have imparted such high energies.

Now in the Space Boundary theory we have a new cosmic event which is in the form of a galaxy formation event where an amount of energy corresponding to a mass equivalent of 2×10^{42} kg is released from the fabric of spacetime.

It is proposed that this galaxy formation event is the source of the cosmic rays of such high energy.

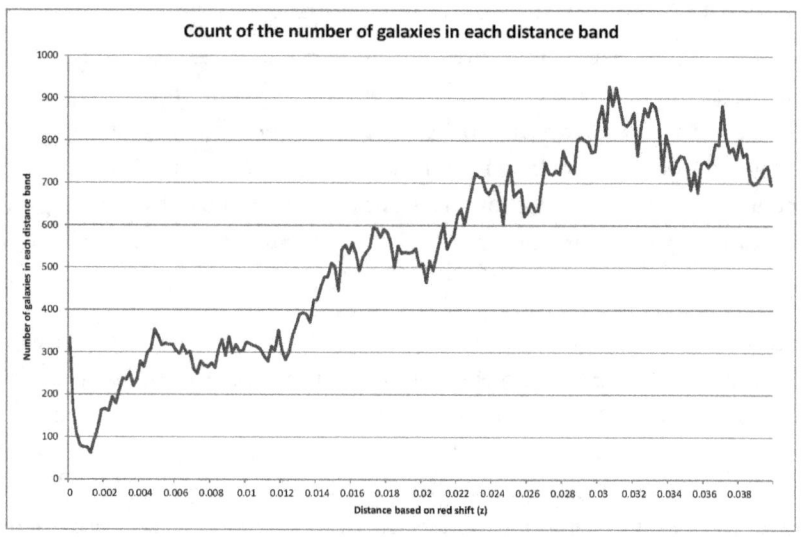

FROM APPENDIX 1: DISTANCE BANDS ARE 2.75 MILLION LIGHT YEARS

The distribution of galaxies out to a distance of 550 million light years is presented in the chart above. We need to establish a model for galaxy formation which yields a pattern which approximates to the above chart.

The proposed model is one in which galaxies form in a universe which is finite with a space boundary at radius R. Galaxies form with the dominant position of galaxy formation moving from the centre of the universe towards the boundary.

Table 14 provides the numerical data to tabulate the position of galaxies. The table uses the calculation that the first galaxy formed 126 billion years ago and the number of galaxies increased by a factor of 20 every 14 billion years.

From the data in Appendix 1 and Appendix 21 we can obtain the current position of the galaxies that formed in the range 0 to 550 million light years.

Table 14 - 1 Galaxy position					
Time Interval BY since first galaxy	0	14	28	42	56
Total Number of Galaxies	1	20	400	8000	160000
Current position of Galaxies (MLY)	9	15	33	148	660
Elapsed time since formation (BY)	126	112	98	84	70

The data of Table 14 -1 is then extended to Table 14 -2 in such a way that the number of galaxies increases by a factor of 20 every 14 billion years and the current position of each additional set of galaxies increases by a factor of 2.

Using a factor of 2 in the model achieves the desired result that there are sufficient galaxies within a radius of 8770 million light years at a time 13.8 billion years ago which is required for the formation of an event horizon.

Table 14 - 2 Galaxy position					
Time Interval BY since first galaxy	70	84	98	112	126
Total Number of Galaxies	20^5	20^6	20^7	20^8	20^9
Current position of Galaxies (MLY)	1320	2640	5280	10560	21120
Observed position of Galaxies (MLY)	1260	2420	4480	8038	12875
Elapsed time since formation (BY)	56	42	28	14	0

The data in Table 14 - 2 gives the estimated current positions of galaxies.

Consider the 64 million galaxies that formed 42 billion years ago and are currently at a distance of 2640 million light years. If we go back in time 2.42 billion years, these galaxies would be located at 2220 million light years and the travel time for light to reach us from this distance is 2.42 billion years taking into account the expansion of space.

The time that light has travelled from this set of galaxies is 2420 million years and this would be the basis of any observational measure of distance. This should be taken into account when checking the model against observational data of number of galaxies by distance.

The method of calculating the observed distance starting from $Y = 2640$ million light years uses $Y = X \exp(K\,T)$ and $X = (1/K)\,(1 - \exp(-\,K\,T)$. In these formulae T is measured in years and X and Y in light years.

$Y = (1/K)(\exp(K\,T) - 1)$

$T = (1/K)(\log(K\,Y) + 1)$

This formula is used to derive the observed position of galaxies in Table 14 - 2.

Appendix 15 Galaxy formation analysis

The formation of galaxies in the universe uses energy derived from the expansion of space and it is proposed that each cubic metre has the same matter forming potential corresponding to around one hydrogen atom per cubic metre. This implies that there is a direct linear relationship between the total mass of galaxies in the universe and the total volume of the universe.

However, the matter formation takes place not in the form of individual hydrogen atoms but in the sudden release of tension in the fabric of spacetime leading to a galaxy formation event.

Suppose the first galaxy formed at time T_F. Then after 14 billion years the radius of the universe will have increased by a factor e (2.718 . . .) and the volume of the universe will have increased by a factor e^3 (approximately 20). This implies that the total number of galaxies in the universe increases by a factor of 20 every 14 billion years.

Recognising that 20^9 is 512 billion and this a reasonable estimate for the total number of galaxies in the universe today then we can estimate that time T_F is around 126 (=14 x 9) billion years ago. Considering the estimated age of the universe is greater than 300 billion years, this implies that there was a period of more than 150 billion years during which time there was no matter formation but just a build up of tension in the fabric of spacetime.

The observational data from Appendix 1 suggests that galaxies formed progressively over time with the position of formation closest to the centre of the universe and moving progressively outwards away from the centre. We can use the current position of a galaxy to estimate the time since that galaxy formed. This is the time of initial formation not the time of subsequent mergers with other galaxies.

If we look at the data of Appendix 1 we find that the number of galaxies in distance band n is approximately 5.5 n. Each distance band is 2.75 million light years (MLY). This means that the number (N) of galaxies currently within a radius R (MLY) is approximately R^2 / 2.75. So for example with R equal to 550 MLY the number of galaxies currently within that radius is 110,000.

We have also concluded above that the number of galaxies formed after a time T billion years (BY) since T_F is 20 ^ (T/14). Putting these results together we get:

$$20 \char94 (T/14) = R^2 / 2.75$$

Using natural logarithms: $(T/14) \log(20) = 2 \log(R) - \log(2.75)$

$T = 9.346 \log(R) - 4.728$ (This estimate is valid in the range R = 20 to R = 550)

When R = 26 MLY, T = 25.75 BY. This implies that the Milky Way galaxy first formed 25.75 billion years after T_F which means that the Milky Way galaxy formed approximately 100 billion years ago.

Appendix 16 Event horizon formation analysis

In the Space Boundary theory, the Cosmic Microwave Background Radiation (CMBR) arises because there is sufficient mass in the universe to create an event horizon at the Schwarzschild radius. The observation of the CMBR shows that it has a look back time of 13.8 billion years.

It has been deduced that the event horizon was located at 8.775584 billion light years at a time 13.8 billion years ago. The CMB radiation started at a distance of 8.775584 billion light years but because of the expansion of space the radiation took 13.8 billion years to arrive.

This uses the formula $14*(1 - \exp(-13.8/14))$ which is $(1/K)(1 - \exp(- K T))$ with T = 13.8 billion years.

The event horizon is located at the Schwarzschild radius which is given by the formula:

$R = 2GM/c^2$

The mass contained within the Schwarzschild radius at a time 13.8 billion years ago is calculated by putting R = 8.775584 billion light years in the Schwarzschild radius equation:

$Rc^2/2G = 55.9 \times 10^{51}$ kg.

Taking a typical galaxy to have a mass of 2×10^{42} kg, this corresponds to 27.95 billion galaxies being located within the event horizon at a time 13.8 billion years ago.

This corresponds to an average matter density of 23.32×10^{-27} kg m^{-3} at a time 13.8 billion years ago. The uniform expansion of the universe over the last 13.8 billion years has been exp(13.8/14) = 2.6797 which corresponds to a volumetric expansion by a factor of 19.2429.

The expected observed average matter density now (23.32/19.2429) should be 1.2119 x 10^{-27} kg m-3 based on the presence of the event horizon. This compares with the estimate of Appendix 4 of 0.373 x 10^{-27} kg m^{-3} for the distribution of galaxies out to 500 million light years.

This suggests that the matter density profile with increasing distance from the centre of the universe starts by decreasing in value and then increases. The CMBR radiation is coming from a distribution of galaxies constrained to be located just inside the event horizon and these contribute to the increase in matter density averaged over the entire region of space within the event horizon.

Appendix 17 Galaxy Recession Velocity and Dark Energy

In the Big Bang model, the universe is considered to be isotropic so the expansion of space must be the same everywhere. Then with the Big Bang theory, the only way of accounting for the discrepancy in the recession velocity of galaxies is to assume that the rate of expansion of space has been lower in the past and this has been explained as caused by Dark Energy.

Moving to a model of the universe in which space is finite with a boundary, this defines a unique frame of reference (the CMB rest frame) and the universe has a centre. Then it is possible to consider a variation in the Hubble constant value with distance from the centre of the universe.

The recession velocity of any galaxy away from the centre of the universe is made up of two components V_E and V_G where V_E is the recession velocity due to the expansion of space and V_G is the velocity of the galaxy through space. The observed recession velocity is $V_E + V_G$ with some adjustment to take into account the fact that our point of observation is moving away from the centre of the universe at 552 km/s.

To illustrate the model let us take the Hubble constant of 72.5 km/s per Mpc for the local region and 67.5 km/s per Mpc for the distant region. Now consider the galaxies currently located at a distance of 2640 million light years which we take to be located in the distant region where the observed recession velocity is 67.5 km/s per Mpc.

The value of V based on 72.5 km/s per Mpc is 58626 km/s and the value of V based on 67.5 km/s per Mpc is 54583 km/s. The value of V_E must be 58626 km/s and the value of V_G is then - 4042 km/s (4042 km/s in the direction towards the centre of the universe). Using observational data it is possible to derive V_G against distance.

This model assumes a uniform expansion of space over time and distance with a value of K equal to 1 / (13.5 billion) light years per year per light year. The value of V_G has to be derived from observational data.

When creating a model to calculate the expected value of V_G as it increases in magnitude with distance it will be better to use the Einstein equations rather than the Newtonian equations because we know that the value of V_G must increase so that $V_E + V_G$ is zero as we approach the event horizon.

The conclusion is that there is no dark energy and the expansion of space is constant over time and distance. The variation in the observed value of the Hubble constant with distance is as a result of the movement of galaxies through space relative to the expanding CMB rest frame.

Appendix 18 Galaxy distribution

There are some general principles for the position of galaxy formation. The formation of a galaxy occurs when the tension in the fabric of spacetime reaches a threshold. The radius of curvature of space is lower for points closer to the centre of the universe so the curvature is greatest here and the potential for release of energy for matter formation is greater.

Once a galaxy is formed, this has the effect of releasing the tension in the fabric of spacetime so that once a region is well populated with galaxies, there will be a tendency for the region of formation of galaxies to move further away from the centre of the universe.

If the majority of galaxy formation activity is taking place at a distance P from the centre of the universe, then we could still have galaxies forming at a distance of 50% of P or even 25% of P but in decreasing numbers. When galaxies form within an existing region of established galaxies, they form at a point in expanding space and adopt the corresponding expansion recession velocity which applies at that distance. The results of Appendix 1 are consistent with this model.

It is also the case that there will be galaxy formation activity taking place beyond the distance P in decreasing quantities as you move towards the boundary of the universe.

An important consideration is that when a galaxy formation event occurs, the release in the tension of the fabric of spacetime propagates outwards from the galaxy at the speed of light. This implies that when a region of space has a higher tension in the fabric of spacetime, there can be several galaxies formed in a cluster without the reduction in tension being an influence in preventing the other galaxies in the cluster from forming.

Also, once a cluster of galaxies has formed, there will be a general reduction in the tension in the fabric of spacetime which moves outwards in an expanding spherical region. This will tend to create a void on the side of the cluster further from the centre of the universe as the region of galaxy formation moves outwards. Galaxy formation can still take place outside the void region and tend to form walls and sheets of galaxies as is observed.

Appendix 19 Elliptical and Spiral Galaxies

A galaxy formation event creates a spherical region of hydrogen gas and di-neutrons. With no other interaction, this leads to a spherical region of stars surrounded by a spherical region of dark matter (di-neutrons). This covers spherical galaxy formation but it is necessary to explain the formation of elliptical and spiral galaxies.

It is proposed that elliptical and spiral galaxies form as a result of the merger of two spherical gas clouds (hydrogen and dark matter). Subsequently spiral galaxies may also merge to produce more complex and irregular forms of galaxy.

Consider two spherical gas clouds of hydrogen and dark matter A and B which formed at approximately the same time with a separation of two million light years. At the time of formation, the gas clouds will have a radial velocity with reference to the centre of the universe and this velocity due to the expansion of space will be proportional to the distance from the centre of the universe.

The expansion of space continues to move the gas clouds A and B further apart so that it is approximately 5.3 billion years before the expansion velocity is matched by the velocity due to the acceleration under the mutual gravitational attraction of A and B. It takes a total of 14.8 billion years before the gas clouds A and B move together and start the merging process.

If you consider the system A and B with no gravitational attraction from other gas clouds, then the relative expansion velocity will be along the line AB and the gravitational acceleration will be along the line AB in the direction tending to pull A and B together. In this case, when the gas clouds merge it will be a direct merger with no angular momentum so an elliptical galaxy is formed.

Taking into account the presence of other galaxies forming in the vicinity of galaxies A and B there will be a corresponding disturbance of the direct path of merger of A and B. A deviation of just 1% in direction has the effect over a distance of two million light years of separating the direction vectors by 20,000 light years. This is sufficient to result in the rotation of the merged galaxy and the formation of the curved spiral arms.

As the gas clouds A and B approach each other, the material furthest away from the centre of the system AB will experience less gravitational acceleration. This has the effect of generating a tidal effect whereby the central mass of the spherical gas cloud leaves a trail of gas which eventually become the spiral arms which will wrap around the galaxy as it rotates.

As the gas clouds merge, the regions close to the centre and in the spiral arms experience an increase in density which then leads to star formation.

The merger includes the dark matter halo of the two galaxies to eventually form a dark matter halo of the spiral galaxy. This dark matter halo helps to preserve the shape of the spiral structure of the galaxy.

There is a further feature of spiral galaxies which needs explanation, namely the barred spiral shape. It is proposed that the central bar of a barred spiral galaxy develops over time so that barred spiral galaxies tend to be older and therefore represent a higher percentage of the total spiral galaxies in the region closest to the centre of the universe where the oldest galaxies are located.

The suggested mechanism for the formation of the bar is that stars in the spiral arms experience a small excess in gravitational acceleration along the direction of the spiral arm. The acceleration of stars towards the centre of the galaxy is balanced by the rotation of the galaxy but the effect of gravitation within the spiral arms tends to result in the stars not yet incorporated into the bar but still in the curved spiral arms being accelerated until they arrive at the point at the end of the bar. Here there is the net deceleration gravitational effect of the trailing spiral arm so the stars take up the position at the end of the bar.

The spherical gas clouds involved in the merger have a structure comprising a central black hole (neutron star with an event horizon) surrounded by a spherical region of gas (with some star formation activity) surrounded by a spherical dark matter halo. The black hole is typically around one thousandth of the mass of the galaxy. The dark matter is typically around five times the mass of the stars and hydrogen gas. The dark matter extends to a radius of around five times the radius of the star region.

At the time of formation of the spherical gas clouds there are no stars yet formed, but there is sufficient time during the 14.8 billion years (of the example above) before the galaxies merge, for some star formation to take place in the hydrogen gas of each galaxy. As the galaxies merge to form a spiral galaxy, the black holes will merge to form a rotating neutron star at the centre of the spiral galaxy. This merged neutron star will have an event horizon and take the form of a black hole. This black hole will have the combined mass of the black holes of the merging galaxies, less any mass lost into energy (gravitational waves) in the merger.

As the galaxies merge, the dark matter will also experience tidal effects so that the spiral arm of stars will be surrounded by a spiral arm of dark matter of around five times the mass and five times the length and width of the visible spiral arms. This will add to (and dominate) the gravitational effect described in the formation of the barred spiral galaxy.

The merger of the hydrogen gas in the two galaxies has the effect of creating a density variation which lead to further star formation in the spiral galaxy after its initial merger.

The merging of galaxies has the effect of creating regions of matter with angular momentum. This implies that any object in the universe with rotation acquired its angular momentum as a result of a galaxy merger.

Rotating neutron stars, black holes, binary and multiple star systems, planetary systems all acquired angular momentum as a result of the host galaxy be formed from a merger of two or more galaxies at an earlier time. The Milky Way galaxy must have formed as a result of a merger of galaxies.

Initially all galaxies form as a spherical region of gas and dark matter with no angular momentum. Such galaxies may then go on to merge to form a spiral galaxy and subsequently spiral galaxies can fall together under gravity to merge again.

Appendix 20 Our position in the universe

In the Space Boundary theory the universe has a spherical boundary and the centre of the universe is defined as the centre of this sphere.

The cosmic microwave background measurements are redshifted in such a way that the Milky Way galaxy appears to be moving relative to the CMBR rest frame at a speed of 552 km/s. The centre of the universe, as defined above, is at rest in the CMBR rest frame.

By analysing the redshift data we can identify the direction of our movement relative to the CMBR rest frame and if we trace back in this direction, the centre of the universe is located somewhere along this path. The distance to the centre of the universe can then be estimated using Hubble's law.

Imagine being an observer at the centre of the universe which is at rest in the CMBR rest frame. We would observe the Milky Way galaxy moving away at 552 km/s. We would conclude that the distance from the centre of the universe to our galaxy is around 7.903 megaparsec (Mpc). This is based on Hubble's law and a Hubble constant value of 69.84 (km/s) / Mpc.

This places our Milky Way galaxy at a distance of 25.77 million light years from the centre of the universe.

The apparent direction of movement of the Milky Way galaxy relative to the CBMR rest frame is towards Galactic Longitude 175.5 and Galactic Latitude -24. If we take the opposite direction (Lon 355.5, Lat 24) and move to a point at a distance of 25.77 million light years then this is the location of the centre of the universe.

From this point at the centre of the universe, taking the data from the NASA Extragalactic Database to determine the position of galaxies it is found that there are no galaxies within 9.2 million light years of the centre.

Appendix 21 Galaxy distribution near the centre of the universe

The positions of all known local galaxies can be obtained from the NASA Extragalactic Database (NED) and in the data you find Longitude (Galactic), Latitude (Galactic), Velocity (km/s) and Redshift. With this data it is possible to do a translation of coordinates so that the positions of each galaxy relative to the centre of the universe is known.

The best way of doing this is in Cartesian coordinates where S = 25.77 million light years (MLY)

X = S cos(Lat) cos (Lon);
Y = S cos(Lat) sin (Lon);
Z = S sin(Lat);

Then the position of the centre of the universe relative to the Milky Way galaxy is: XC = 24.98 MLY; YC = - 1.9 MLY; ZC = 10.8 MLY

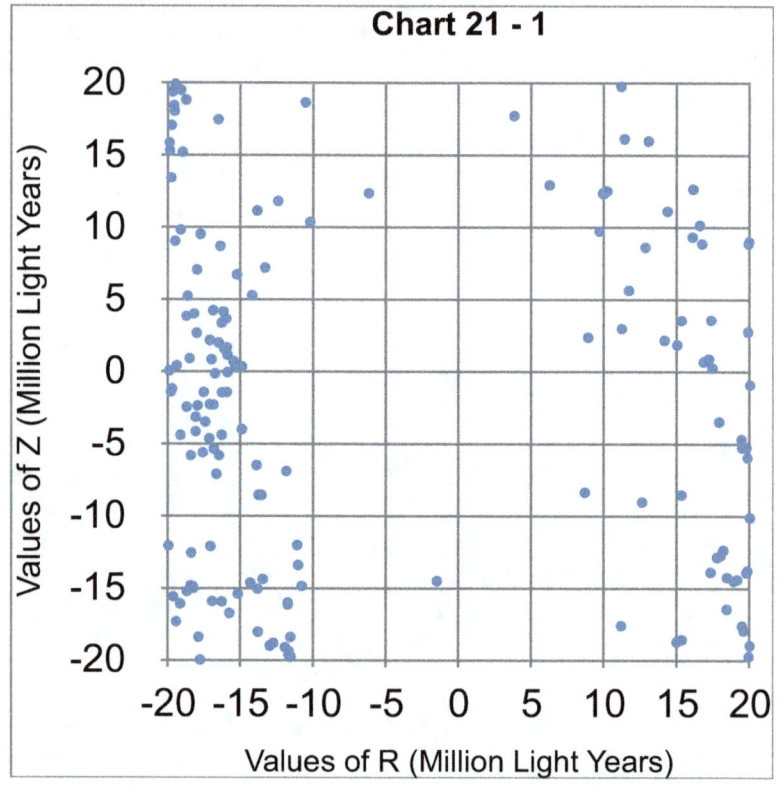

Then knowing the position of any galaxy relative to the Milky Way galaxy (XG, YG, ZG) you can calculate the position of the galaxy relative to the centre of the universe: (XG - XC, YG - YC, ZG- ZC).

This can then be plotted as in Chart 21 - 1. The vertical axis is the Z axis and the horizontal axis of the plot is the values of R where $R^2 = X^2 + Y^2$. The view is along the X axis and positive values of Y are on the right and negative values of Y are on the left. This was done to illustrate the point that there is a spherical region around the centre of the universe of radius 9.2 million light years where there are no galaxies. This is expected because all galaxies have been moving away from the centre of the universe for billions of years.

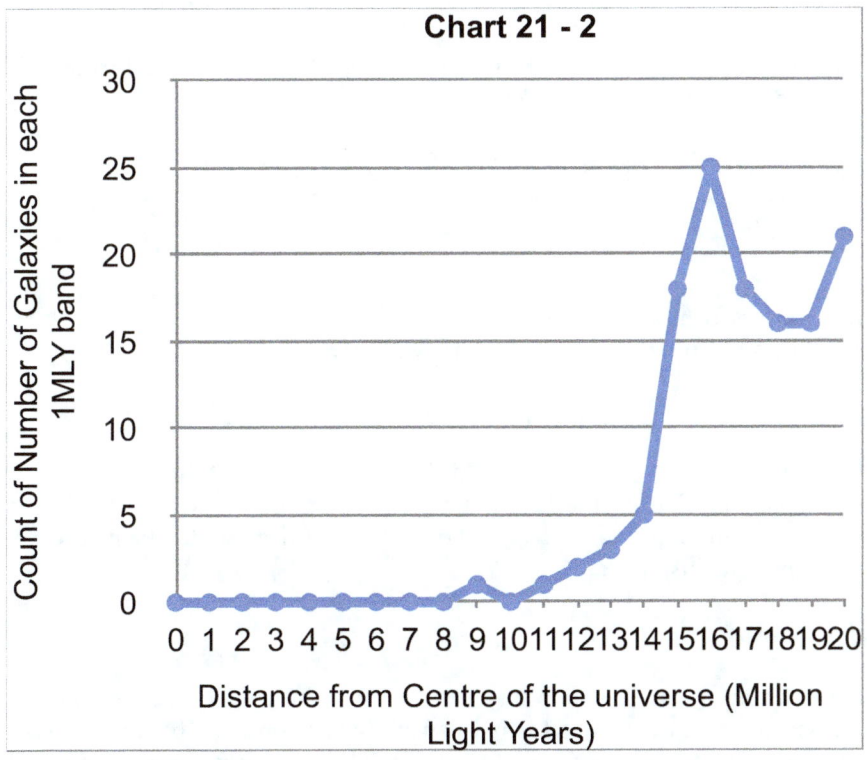

The plot in chart 21 - 2 reveals the initial galaxy formation activity following the formation of the first galaxy 126 billion years ago. There is a steady initial increase in the number of galaxies forming and it reaches an initial peak of 25 galaxies and then drops down again before starting to increase again.

This peaking effect is similar to the pattern of Appendix 1. It is proposed that the new galaxy formation activity is affected by the previous galaxy formation activity which reduces the tension in the fabric of spacetime for a period resulting in fewer galaxies being formed.

[In a revised model of galaxy formation, distance from the centre is not a good indication of the age of the galaxy. The galaxy shown at 9 million light years is not the first galaxy to form. The distribution of galaxies is built up over an extended period of time. However, the number of galaxies by distance at the end of the formation process represents the available energy release from the expansion of curved space.]

Appendix 22 Schwarzschild Radius and Matter Density

The Schwarzschild radius formula can be combined with the formula for matter density in the following way:

The Schwarzschild radius formula is: $R = 2\,G\,M\,/\,c^2$

The formula for matter density is: $D = M\,/\,(4/3\,\pi\,R^3)$

So $R = 2\,G\,D\,(4/3\,\pi\,R^3)\,/\,c^2$

So for a uniform spherical distribution of matter of density D, we can see that an event horizon will form at radius R where:

$$R^2 = 3\,c^2\,/\,(8\,\pi\,G\,D) = Q\,/\,D$$

The constant $Q = 3\,c^2\,/\,(8\,\pi\,G)$ has the value: 1.6073745×10^{26} kg m^{-1}

The lower the density the larger the value of R. For very low density the event horizon is at a large radius. For very high density the event horizon has a small radius.

High density case

Consider the maximum possible density of matter which corresponds to the density of a neutron. This is calculated as 1.598695×10^{18} kg m^{-3} based on a neutron radius of 0.630058×10^{-15} m Then the Schwarzschild radius R is 10.027 km which corresponds to the smallest possible black hole at around 3.4 solar masses. We know this is the smallest possible black hole because we have chosen the maximum possible density. This is in agreement with observations so we must conclude that any greater matter density is not possible and matter does not collapse to a singularity.

Intermediate density case

To visualise an intermediate density, consider a mass distribution with density of one gram per cubic metre. $D = 1 \times 10^{-3}$ kg m^{-3}. Then for a mass distribution of this density with a radius of 400 billion km (0.0424 light years) this would form an event horizon at the surface of the spherical region containing the mass distribution.

It is clear from this example that we have an event horizon without a spacetime singularity. An observer outside this distribution of mass would detect a black hole with a mass of around 0.27×10^{42} kg which is around one tenth of the total mass of a typical galaxy.

Low density case

For the low density case, consider a large region of space with a matter density of one hydrogen atom per cubic metre. This corresponds to a density of 1.67355×10^{-27} kg m-3. Now we know that any event horizon that would form will be at a great distance so we need to consider the density of the currently observed material as it would have been in the past.

If we go back in time by 13.8 billion years with the assumption of a uniform expansion of space K then the linear expansion over that period would have been exp (K T) which is 2.6797. Going back in time the density would be increased by the cube of this value which is: 19.2429. So if we observe a matter density today of one hydrogen atom per m3 then 13.8 billion years ago the density of this same body of matter would have been 3.22×10^{-26} kg m^{-3}. A matter density of 3.22×10^{-26} kg m^{-3} will result in an event horizon located at a distance of 0.70648×10^{26} m which is 7.46759 billion light years.

We need to consider the transit time for radiation which started off at a distance of 7.46759 billion light years when the radiation was emitted. We have previously calculated the relationship between the look back time T and the original distance of separation S as $S = (1/K) (1 - \exp(- K T))$.

$$T = (1/K) \log (1 / (1 - K S))$$

This would imply that we would expect to see radiation with a look back time of 10.67194 billion years.

Actually what we observe is radiation with a look back time of 13.8 billion years. This is the Cosmic Microwave Background radiation. If we change some of the assumptions for example the matter density that we observe is just 0.724 hydrogen atoms per cubic metre, then with the lower density, we get a larger Schwarzschild radius and the look back time is then 13.8 billion years.

So the conclusion from the analysis of the Schwarzschild Radius formula using matter density is that:

1) Black holes must be greater that 3.4 solar masses (as is observed) and do not contain a singularity

2) The Cosmic Microwave Background Radiation is radiation coming from a source located just within the event horizon.

Appendix 23 Expansion Factor K

The expansion factor K used in the earlier calculations has been 1 / (14 billion) light years per year per light year which corresponds to an H_0 value of 69.84 km/s per Mpc. If we instead take and expansion factor K value of 1 / (13.5 billion) light years per year per light year this gives an H_0 value of 72.435 km/s per Mpc.

It is useful to explore the implications of this higher value of the expansion of space on the calculations completed so far.

The revised calculations with K = 1 / (13.5 billion) are:

Time since the formation of the first galaxy: 121.5 billion years

Event horizon at a time 13.8 billion years ago: 8.64277 billion light years

Density within the event horizon 13.8 billion years ago: 24.0452 x 10^{-27} kg m^{-3}

Volume expansion factor over the last 13.8 billion years: 21.47

Expected currently observable matter density: 1.12 x 10^{-27} kg m^{-3}

Age of the universe - at least: 322 billion years

Position of Milky Way based on 552 km/s: 24.857 million light years from the centre

If the position of the Milky Way galaxy is taken to be 25.78 million light years from the centre (based on observation of the galaxy distribution close to the centre) then this leads to a value of V_E (based on H_0 = 72.435) of 572.5 km/s.

With the observed velocity of the Milky Way relative to the CMB rest frame given by $V_E + V_G$ = 552 km/s this gives a radial component V_G of - 20.5 km/s (20.5 km/s towards the centre of the universe)

To calculate the age of the Milky Way we use the formula $20 \char`\^ (T/13.5) = R^2 / 2.75$

This converts to T = 9 log (R) - 4.558 which (for R = 25.78 MLY) means that the Milky Way galaxy formed 24.68 billion years after the first galaxy formed 121.5 billion years ago.

This puts the age of the Milky Way galaxy at around 97 billion years.

[Distance from the centre can not be used as an indication of the age of the galaxy. Revised estimate of the age of the Milky Way is 56 to 70 billion years.]

The revised calculations have not been applied throughout this paper because there is still some uncertainty over the precise value of H_0 from astronomical observations. This section of the paper shows the sensitivity of the calculated results to adjustment of H_0.

6 RESOURCES

These YouTube videos are located in the channel @OpenWorldRichard with the name Richard Lewis and are freely available to watch.

1. The Explanation for Dark Matter and Dark Energy

This is the video presentation presented in chapter 2 of this book.

2. The Formation of the Solar System

This is the video presentation presented in chapter 3 of this book.

3. Galaxy Distribution

The presentation covers the results of an analysis of galaxy positions based on data downloaded from the NASA Extragalactic Database (NED). The positions and red-shift of approximately 110,000 galaxies have been taken and these galaxies are located at a distance of up to 550 million light years. The galaxies have been grouped into distance bands corresponding to steps of 2.75 million light years (red shift $z = 0.0002$) and the number of galaxies in each distance band is counted. The results of the analysis show that the progress of the galaxies counts with distance does not at all correspond to what one would expect from the Big Bang theory.

The galaxies seem to have formed as part of the progress of galaxy formation events taking place at increasing distance. If the galaxies had formed from a widespread cloud of gas and dust throughout the universe then one would expect much more randomness in the galaxy counts due to statistical variation in the process so that successive distance bands vary by typically plus or minus 50% or more. This is not what is observed. Instead there is a steady progress of the number of galaxies in a regular pattern. In another video referenced within the actual angular positions of individual galaxies is presented for each distance band. This gives a visual representation of the evolution of galaxy positions and the sense of movement observed is due to the progressive adjustment of positions in each distance band.

Finally a plot is provided of galaxy positions with reference to the centre of the universe. This reveals a void of galaxies out to 9 million light years. A related plot shows the number of galaxies by distance from the centre of the universe.

4.   Andromeda

The video starts with the analogy of observing a falling ball through a window and from a precise measurement of the speed being able to calculate the height from which it was dropped and the time when it was dropped. Similarly we know that the Andromeda galaxy is moving towards the Milky Way galaxy at 110 km/s and is at a distance of 2.5 million light years. Starting from a distance of separation of X light years where the galaxies are at rest in expanding space the initial recession velocity is given by Hubble's law.

Then the effect of the expansion of space and gravitational acceleration is calculated using spreadsheet where each row in the spreadsheet corresponds to an elapsed time. From this calculation using different starting values of X it is shown that the currently observed distance and velocity of approach must correspond to an initial separation of 2.44 million light years, followed by a maximum separation of nearly 5 million light years and a total elapsed time to reach the currently observed position and velocity of 53.9 billion years.

The results on sheet 2 also show that for an initial separation of 2 million light years the separation never reaches 2.5 million light years and for any initial separation of greater than 2.46 million light years the direction of travel is never reversed. This illustrates the interplay between the expansion of space and gravity and helps to explain the stretching of two spherical gas clouds to form the two spiral arms of a galaxy during an initial merger of galaxies.

5. Spiral Galaxies

If you have watched the video on the past motion of the galaxy Andromeda you will remember the interplay between the expansion of space and the acceleration due to gravity. If you then think of two spherical regions of gas an dust with a radius of 0.5 million light years and a distance of separation of 2 million light years you will see that the spherical shape becomes distorted and stretched so that in the resulting merger the remote sides of the sphere become the two spiral arms of a symmetrical spiral galaxy. This explains the formation of the spiral arms and the origin of the angular momentum apparent in the rotation of the galaxy

6. How Far Away

This presentation addresses the problem of determining how far away an object was located at the time that the radiation was emitted. When measurements are made of luminosity and redshift the observation depends on the time taken for the light to travel. This is because both the luminosity and the red shift values which we measure depend on the actual distance travelled through expanding space. From these observations we can calculate how far away the object was located when the light was emitted.

Using the assumption that the expansion of space is uniform and based on an expansion rate of 1 part in 14 billion light years per year per light year (69.84 km/s per Mpc) it is shown that an observation of radiation with a travel time of 13.8 billion years corresponds to a radiation source which was at a distance of 8.77 billion light years at a time 13.8 billion years ago. This supports the hypothesis that the Cosmic Microwave Background Radiation is radiation coming from a source which was located just inside the event horizon of the universe which was located at a radius of 8.77 billion light years at a time 13.8 billion years ago. The position of this event horizon is consistent with the estimated matter density within the event horizon of the universe at that time.

7. Black Holes

The presentation explains the theory that all black holes contain a neutron star of the corresponding mass. The theory discards the idea that black holes contain a singularity at the centre of the black hole. Also the idea of black hole evaporation is discarded. The formation of an event horizon is considered for low, intermediate and high densities. The presentation rejects the hypothesis that black holes contain a singularity.

8. Galaxy Formation

A mathematical model of galaxy formation is presented to explain the distribution patterns of galaxies.

7 OPEN WORLD BOOK SERIES

Open World book 1: The Evolution of the Universe.

Presents a summary of the evolution of the universe which is finite with a space boundary. Galaxy formation takes place in individual galaxy formation events with energy for matter formation coming from expanding curved space. Galaxy formation started close to the centre of the universe at a time around 126 billion years ago and the position of subsequent galaxy formation moved away from the centre.

Open World book 2: The nature of matter.

Light travels as a wave in the medium of space. Matter particles such as the neutron proton and electron are looped waves in the medium of space. Starting from this wave oriented view the results of quantum theory and particle physics are explained.

Open World book 3: The Conscious Brain

Consciousness is defined as the subjective experience that we have from the operation of our brain. From this definition the importance of "Focus of Attention" is highlighted and a neuron network cause is described.

Open World book 4: The Global Economic System

An enhancement to the Global Economic System is proposed which is aimed at easing deprivation. It also provides incentives towards peaceful coexistence between nations of the world.

Open World book 5: Openness

The importance of guiding principles is identified. "I belong to a world in which all may live in peace following the principles of honesty trust and mutual respect. Honesty without fear; Trust filled with hope; and mutual respect for every individual person"

ABOUT THE AUTHOR

I was born in the UK in 1946 and educated at Forest School and Cambridge University (St Catharine's College) where I read mathematics between 1964 and 1967.

I worked for the Marconi company for two years completing the graduate training program and then I worked on real time programming on the Myriad computer.

In 1969 I joined Northern Electric in Toronto Canada working on the real time programming of the No1 ESS telephone exchange. I joined Bell Northern Research in Ottawa in 1971 soon after its formation from Northern Electric and Bell Canada to work on the SL-1 digital telephone exchange where I led the software development for that product.

Returning to the UK in 1977, I worked for Nortel providing technical support for sales in Europe and the Middle East. I worked for three years in the city of London for Kleinwort Benson in the department responsible for providing telephone and data communications to the group.

I have always taken a keen interest in theoretical physics and cosmology and I have made use of published research on the internet, particularly from NASA (NED) and Wikipedia.

The papers on Academia have been published since retirement and I have obtained useful information from books and YouTube videos which describe the unexplained issues of physics and cosmology.

I am now working on Open World which seeks to answer some fundamental questions: How did the universe evolve? What is the fundamental nature of matter? How does the brain work? How should society organise economically? What are the fundamental ethical principles? How can we achieve personal, national and international peace?

www.ingramcontent.com/pod-product-compliance
Lightning Source LLC
Chambersburg PA
CBHW062344290526
45794CB00005B/2099